APPLIED
SURVEILLANCE
PHOTOGRAPHY

APPLIED SURVEILLANCE PHOTOGRAPHY

By

RAYMOND P. SILJANDER

A.S., Law Enforcement
Graduate, Institute of Applied Science
Member, Minnesota Division of
International Association for Identification
Formerly, President, Researchco Investigative Services, Inc.
Presently, Police Department, Marshall, Minnesota

CHARLES C THOMAS · PUBLISHER
Springfield · Illinois · U.S.A.

Published and Distributed Throughout the World by
CHARLES C THOMAS • PUBLISHER
Bannerstone House
301-327 East Lawrence Avenue, Springfield, Illinois, U.S.A.

© *1975, by* CHARLES C THOMAS • PUBLISHER

ISBN 0-398-03376-5

Library of Congress Catalog Card Number: 74-234-71

*With THOMAS BOOKS careful attention is given to all details of
manufacturing and design. It is the Publisher's desire to present books that
are satisfactory as to their physical qualities and artistic possibilities and
appropriate for their particular use. THOMAS BOOKS will be true to those
laws of quality that assure a good name and good will.*

Printed in the United States of America
C-1

Library of Congress Cataloging in Publication Data

Siljander, Raymond P
    Applied surveillance photography.

    1. Photography, Legal. I. Title.
TR822.S54   778.9'9'364128   74-23471
ISBN 0-398-03376-5

To
Dwaine Raymond

INTRODUCTION

T HE AUTHOR, in writing this book, has attempted to fulfill a two-fold need of investigative agencies. While it is true that there is available on the market a piece of equipment to satisfy almost every surveillance photography need one may encounter, there are two main obstacles. The first obstacle is that much of the specialized photographic equipment on the market is prohibitively expensive. The second obstacle is the lack of trained personnel to effectively use such equipment. It is the author's belief, however, that the average law enforcement officer or investigator with a background in general photography can master the techniques set forth here if he employs some thought and takes the time to engage in some experimentation.

Every effort has been made to show where shortcuts can be effected to save considerable sums of money on equipment and still be able to obtain the photographic results needed. A few pieces of equipment, although somewhat expensive, have been introduced in rather general terms for familiarization purposes. Except for a few specific areas, an attempt has been made to put forth information in general terms that can be applied to most all 16mm motion picture cameras and 35mm single lens reflex cameras. Any deviation from this is coincidental and done only to help clarify some specific point. Also, the techniques discussed herein are those that will generally apply to manned equipment rather than nonmanned equipment.

It is not the intent of this book to go into the various aspects of darkroom work or to get into the area of general photography. Rather, it is intended to be specialized in nature, and consequently, to fully benefit from this writing, one must have at least a basic knowledge and background in photography and darkroom techniques. Finally, the legal aspects of photography are discussed only briefly as there are current writings available that deal specifically with that topic.

ACKNOWLEDGMENTS

IT IS A PLEASURE to acknowledge and officially thank a number of people for generous and helpful cooperation, before and during the task of writing this book. The author's heartfelt thanks and appreciation goes out to Sgt. Larry Hauck, New Brighton Police Department, New Brighton, Minnesota; Sgt. Jerry Wallin, Blaine Police Department, Blaine, Minnesota; Frank Agar, college photography instructor and free-lance photographer, Minneapolis, Minnesota; Gary Rasmusson, Sly-Fox Films, Inc., Minneapolis, Minnesota; Howie Normandin, Century Communications, Minneapolis, Minnesota; David Roston, attorney, Minneapolis, Minnesota; lifelong friend John E. Waaraniemi; friend, Roy Haglund, who stimulated a very early interest in photography; wife, Ruth; and sisters, Detta and Yvonne.

R. P. S.

CONTENTS

Page

Introduction vii

Acknowledgments ix

Chapter

ONE   CAMERA TYPES GENERALLY USED FOR SURVEILLANCE
PHOTOGRAPHY 3

TWO   TELEPHOTO PHOTOGRAPHY 17

THREE   BLACK AND WHITE FILM VS. COLOR FILM 48

FOUR   SURVEILLANCE PHOTOGRAPHY AT NIGHT USING ULTRA-
HIGH SPEED FILMS 49

FIVE   PUSH-PROCESSING PHOTOGRAPHIC FILM 58

SIX   INFRARED SURVEILLANCE PHOTOGRAPHY 64

SEVEN   INFRARED SURVEILLANCE PHOTOGRAPHY, 16 MM
MOTION PICTURES 77

EIGHT   PHOTOGRAPHY AT NIGHT USING STARLIGHT SCOPES . 80

NINE   HOW TO OPENLY PHOTOGRAPH A PERSON WITHOUT
HIS KNOWLEDGE 89

TEN   VANTAGE POINTS 92

ELEVEN   SURVEILLANCE PHOTOGRAPHS AND THE LAW . . . 94

Index 99

APPLIED
SURVEILLANCE
PHOTOGRAPHY

CHAPTER ONE

CAMERA TYPES GENERALLY USED FOR SURVEILLANCE PHOTOGRAPHY

A S A GENERAL RULE it will be found that the cameras best suited for surveillance photography are 35mm single lens reflex cameras, 16mm motion picture cameras, and in some unique situations the subminiature cameras that are about the size of a pack of cigarettes or smaller. In an effort to maintain clarity, each will be discussed separately.

35mm SINGLE LENS REFLEX CAMERAS

There is no still camera that enjoys the wide variety of fast lenses that is enjoyed by the 35mm single lens reflex camera. The need for fast lenses arises from the fact that surveillance photography is often under lighting conditions that are less than ideal. Also, because surveillance photography is often under conditions that necessitate the use of moderate to extreme telephoto lenses, a range finder camera is of limited value as there would be no easy way to ensure that the subject is properly framed. Therefore it is the single lens reflex that is the logical choice for surveillance photography. Specific attention will be given to telephoto lenses and their use in a chapter devoted to that topic. There is also a very wide variety of film types that come in the 35mm size. These vary from several types of color film to a number of black and white films as well as several types of special film such as infrared and ultrahigh speed films for specialized applications of photography. Many of the film types that come in the 35mm size do not come in sizes for many of the larger format cameras. Thirty-five millimeter cameras are small and easily transported, and are much easier to work with surreptitiously than larger format cameras. The design of these cameras makes it easy to make rapid sequence exposures because one sweep of the film advance lever advances the film all the way and

3

cocks the shutter making the camera ready for the next exposure. In addition to this, many of these cameras can be obtained
with motor drives that enable one to make sequence exposures
at a rate of three to five frames per second or any number of
slower rates as may be preselected. It is possible with many
brands of 35mm cameras to attach a special back to the camera
that allows one to use more than just thirty-six-exposure rolls of
film. Two-hundred-fifty-exposure backs are common, and Nikon
offers an 800-exposure back that takes a standard 100-foot roll
of film. It is also possible to set up a camera with a motor drive
and make exposures by using a radio control device as far as a
half-mile from the camera.

16mm MOTION PICTURE CAMERAS
General Considerations

As is the case with 35mm single lens reflex cameras, the 16mm
motion picture cameras have a wide variety of lenses, zoom and
telephoto, available to them. In many cases it is possible to use a
C-mount adapter to enable one to use a 35mm camera lens system on the 16mm camera. The C-mount adapter is simply an
adapter that is screwed onto the motion picture camera in place
of the camera lenses. The adapter then allows the 35mm camera
lenses to be mounted to the motion picture camera in the same
manner they would be mounted to the body of the 35mm camera. The big advantage of this is allowing one to take advantage
of the many fast lenses made for 35mm cameras and also to
have a complete lens system for both cameras without having to
buy two lens systems.

There is also available for the 16mm motion picture cameras
a wide variety of film types from which to choose. There are
some activities that can be photographically recorded with a motion picture camera that could not be secured with a still camera.
This would include, but would not be limited to, such things as
a subject's mannerisms, the way a person walks, interaction between two or more subjects, and so forth. Also, many motion
picture cameras are equipped with a single-frame release allowing a person to take sequence shots rather than a straight motion

picture when the need arises. Although this often puts a mechanical strain on the camera shutter mechanism, many motion picture cameras can be set up with an electric sequence unit that allows the camera to take a photo at preselected intervals just as was mentioned concerning the 35mm cameras. If a 16mm motion picture camera with a 100-foot roll of film was set up to make an exposure every seven seconds, it would last about eight hours and make about 4000 exposures. Also possible with the motion picture camera at the opposite extreme is its ability to photograph in slow motion. If the subject is filmed at anywhere between 32 and 48 or 64 frames per second and viewed at about sixteen frames per second, details concerning the subject's movements, the way he walks, and so on, may be critically studied.

Figure 1. The Bolex Rex 5 is a 16mm self-threading reflex camera. As can be seen, the Rex 5 is equipped with a three-lens turret.

Also worthy of consideration is whether or not a camera accepts roll film or a cassette. The cassette, while being more costly, has the advantage over the former in that when filming in the field, rapid film changes are possible. Although the cassette cameras are no longer being built, there are still some used ones available. There are also motion picture cameras that, while they do not accept a cassette, are self-threading thus often making the operation easier and faster than having to do the entire operation oneself. The Bolex Rex 5® shown in Figure 1 is self-threading.

Motion Picture Cameras for Night Work: Three Basic Designs

Most but not all motion picture cameras are designed similar to the three designs that will be discussed in very general non-technical terms. This brief discussion is intended to show why some systems are more desirable than others when it comes to work in low light levels.

Most Super 8 reflex movie cameras are intended primarily for amateur use and an effort has been made to make them very automated and simple to use and yet relatively inexpensive. As a result, these cameras tend to waste a lot of light because of the number of optical elements and beam splitter prisms the light encounters before it reaches the film plane. Let us follow a beam of light from the time it enters the primary lens, which in most cases is a zoom lens, until it reaches the film plane. The primary lens will reduce the light value by about 10 to 20 percent. The light, after passing through the primary lens, comes to a beam splitter that will further reduce the light by about 20 percent when taking part of it for the image that will appear in the viewfinder. We have now lost between 30 and 40 percent of our original light value. As the light continues from the beam splitter it passes through a master lens that will absorb about 5 percent of the light. Next it goes through another beam splitter prism to give some light to the cell of the light meter. Our original light value has now been reduced to less than half, or more than one f-stop. For normal daytime photographing this kind of a system is fine, but it has limited usefulness when it comes to night work using available light.

The second system we will discuss is very effective insofar as maximum light transmission to the film plane goes, but it is also quite expensive. This system is found mostly on 16mm motion picture cameras. The only Super 8 camera utilizing this design is the Beaulieu®, which is a very expensive camera. Unlike all the other Super 8 cameras, the Beaulieu utilizes the C-mount, thus making it possible to utilize a 35mm SLR lens system with it. Getting back now to this basic design, it is a reciprocating mirror-shutter system. Quite simply, it is a shutter that reciprocates up and down. In the "up" position a hole in the shutter lines up with the lens and film format and allows the light to expose the film. In the "down" position a 45-degree mirror that is mounted to the shutter is positioned in place of the shutter opening and directs the light to the viewfinder and light meter instead of the

Figure 2. The Bell & Howell Series 70 camera is a 16mm range finder camera. It too is equipped with a three-lens turret but, unlike the Bolex Rex 5, is not self-threading. The range finder is equipped with a parallax adjustment.

film. This system offers an image in the viewfinder that, while being bright, flutters. The flutter, however, presents no problem. With this system the only light entering the lens that does not reach the film when an exposure is being made is whatever light the lens elements absorb.

The third basic design is cameras with separate viewfinders. The Bell & Howell camera shown in Figure 2 is of this design. The only thing between the lens and the film plane is a shutter; thus, all the light entering the lens reaches the film except for whatever amount is absorbed by the lens elements. Also, with this kind of system the image in the viewfinder is bright. Many of these systems also have a separate window to gather light for the light meter so as not to reduce the amount of light reaching the film.

Except for the reciprocating mirror-shutter that was discussed, most motion picture cameras have a shutter that looks like a disc with a large notch cut out of it. The disc spins and every time the notch passes over the film format, the film is exposed. It used to be that cameras commonly had a notch that was rather narrow. The Bell & Howell Series 70® camera shown in Figure 2 has a 204-degree shutter (the notch is 204°). A recent advancement or improvement is the making of a shutter that has a 230-degree section removed. It can be seen that with the shutter revolving at a given speed, the exposure time will be increased if a wider notch has been made. As stated, the Bell & Howell camera in Figure 2 has a 204-degree opening. When this camera is run at eight frames per second (fps) the exposure time is $\frac{1}{44}$ second. While a camera with a 230-degree shutter will probably not be capable of an exposure time longer than this, it can achieve this exposure time while running at a higher number of frames per second. The advantage of this is the fact that the subject's movements, and so forth, will be recorded more smoothly, thus avoiding the "Keystone Cop" effect. The reader should also bear in mind that most projectors will not run at eight frames per second. Filming the subject with an exposure time of less than $\frac{1}{44}$ second would not be practical as there would be severe problems with image blur, and this is bad enough at ¼₄ second. As with so many other things, this is a game of compromise. One may pay anywhere from several hundred to several thousand dollars for a new late model camera with many of the latest and most advanced features or he may pay $200 for a simpler and perhaps used camera that, while it will do the job, forces one to accept some compromise. It depends a great deal upon one's needs and the budget with which he must work.

In conclusion, when selecting a motion picture camera for work under available light at night, the main points to be taken into consideration and looked for in a camera are fast lenses, slow shutter speeds and the degree of the shutter, and a lack of beam splitter prisms, and so on, that tend to greatly reduce the amount of light that reaches the film. Also worth keeping in mind is how readily one can utilize a C-mount adapter to enable him to use the lens system of a 35mm camera. One should remember also that a large zoom lens with its many elements will tend to rob him of more light than will a fixed focal length lens with fewer elements. Also, the cost of the camera selected must be in keeping with one's budget. The idea is not to buy the best thing money can buy but to buy the best thing money will buy relative to the purchaser's needs.

16mm Films for Night Work

There are several films worth exploring for this kind of work. There is Kodak's 2475 High-Speed Recording® film which can be exposed at E.I.4000. This film, however, is a negative film and consequently it will have to be printed in order to have a positive image. While it is doubtful that one would lose any more through the printing process than he loses when making paper prints in the dark room, the largest concern is the added expense. Also worth trying is 4-X Reversal film with an ASA rating of 320 under Tungsten illumination. This film can be push-processed by two f-stops and will yield very good results. An extra charge is usually required for films that are push-processed and some labs have a limit as to how many stops they are willing to push film; others will push it as far as one wants them to go and simply charge accordingly. It is advisable to contact the lab and discuss this with them prior to exposing a film. One may also want to experiment with Gafpan High-Speed Reversal® film, ASA 500 under Tungsten. It responds quite well to push-processing to E.I.4000. It is rather grainy but the grain does not seem to interfere much with subject identification. If color is essential, consider Kodak Ektachrome EF® or some similar film. This film can be push-processed to E.I.1000 or more.

Because of the number of films presently on the market, and because of the fact that new films are often introduced on the market, one would do well to see a dealer and discuss film needs. The reader is in no way limited to the few film types that have been discussed here.

16mm vs. Super 8 Motion Picture Cameras

Super 8 motion picture cameras are generally not considered to be good surveillance cameras because they have a very small image area, they do not have a wide variety of films available to them, and with the exception of only the most expensive models, one must settle for the lens (in most cases a short-range zoom) that is factory-mounted to the camera body. All these factors tend to make them less effective when it comes to obtaining good subject identification under anything but ideal conditions. A possible exception to this might be when there is a need to film a person inside a building and there is no way to do it but to go into the building with the camera.

There has been a recent trend to make available light Super 8 motion picture cameras for the hobbyist. These cameras have a wide shutter, accept fast film (Ektachrome ASA 160) and have very fast lenses. Kodak has such a camera with an f1.2 lens and Keystone has a camera with an f1.1 lens. Another positive factor of these cameras is that they are relatively small. If in a unique situation and if there were no other way to do it, it would not be difficult to mount one of these cameras in an attaché case or some ordinary-looking package with a one-inch hole in the end of it and an inconspicuous cable release to operate the camera. A female investigator could easily mount the camera in a purse. Such a setup would make it possible to follow a subject into a restaurant, etc., and film him and his actions therein if that were necessary. The fact that the Super 8 motion picture cameras are generally not considered to be good surveillance cameras does not mean that one would be doing wrong to give them a try if he does not have a 16mm camera at his disposal and he has an opportunity to document activity or secure evidence by use of this medium.

Reflex Viewing vs. Range Finder Cameras for Telephoto Work

One of the first differences the reader will notice between these two basic types of motion picture cameras is the cost factor, the reflex cameras being considerably more expensive than the range finder cameras. While it is true that the reflex cameras have many advantages over the latter for telephoto work, the range finder cameras, such as the Bell & Howell in Figure 2 are acceptable for this purpose until one begins to get into the area of "extreme" telephoto work. This is where the line is drawn that tends to separate the two. This does not mean that the many departments that have Bell & Howell Series 70 cameras will be severely limited when using them for telephoto work, but if one has the option of choosing between the two types his first choice in almost all cases would be the reflex camera. Unless one gets into the area of "extreme" telephoto work, the range finder cameras should offer little in the way of difficulties.

Shown with the Bell & Howell camera in Figure 3 is a 135mm lens with a 2X tele-extender giving an effective focal length of 270mm. The range finder is for a 150mm lens. To visualize what portion of the viewfinder would be covered by a lens twice as long would be no trick. However, if the subject image is so large that there is a danger of important portions of the scene being left out, the use of that long focal length is unjustified and a shorter focal length should be selected. A focal length of 270 mm on a 16mm camera is about 10½ times as strong as its normal lens of one inch (25.4mm). This will prove to be sufficient for most situations requiring long lenses. Consider this against the

Figure 3. Illustrated is a Bell & Howell equipped with a "C" mount adapter, a 2X tele-extender and a 135mm lens. The effective focal length is 270mm which offers an image magnification of about 10½ times over the one-inch (normal) lens.

fact that a 500mm lens on a 35mm camera offers a magnification ten times over the normal 50mm lens. Finally, one of the other advantages of a reflex camera over the range finder is the problem of parallax when close-up work is done; however, this presents no problem in telephoto work. Parallax basically is the apparent change in the subject position due to a change in the viewer's position. When the film plane is exposed to the subject through one lens and the photographer sees the subject through a second lens system, the factor of parallax is introduced.

An effort has not been made to discuss specific details concerning the many makes and models of motion picture cameras from which one can choose. An effort has been made however to provide enough general background data to allow one to make good decisions and a favorable choice when shopping for a motion picture camera to fulfill his specific needs in surveillance photography.

SUBMINIATURE STILL CAMERAS

While the results that can be achieved by the use of a subminiature camera will in no case equal those possible with a good full frame 35mm camera, it is these little gems that enable investigators, especially undercover agents, to secure photographic evidence that could not be obtained by using a larger and more conspicuous camera or when conditions prohibit working from a distance with telephoto lenses. The cameras being referred to are full working cameras as small, and many smaller, than a pack of cigarettes. Many have in addition to a full shutter speed and aperture range, an internal light meter, several of them being fully automatic. While some have better lenses than others, most of them have multi-element lenses and will perform very well. The price of these cameras vary anywhere from forty to several hundred dollars. Some models, the Tessina® for example, have a spring-wound motor drive that makes it possible to take a series of exposures in rapid succession. The film size of these cameras is quite small and very large blowups are generally not possible. In most cases a five-by-seven-inch enlargement is the maximum. Beyond that point, the grain becomes a limiting factor. Dust spots also become a very real problem due to the degree of enlargement necessary for even a five-by-seven-inch print.

The most common film sizes one will run into when looking at subminiatures are 9.5mm used in the Minox® and Yashicas®, 16mm used by Minolta® and Kodak's Instamatic 110® pocket cameras, and finally, the Tessina which takes 35mm film but does not accept the standard 35mm cassette. The Tessina has its own special cassette. The image size produced by the Tessina is only about 20 x 14mm, not 36 x 24mm as found in a full frame 35mm camera. The lens in the Tessina is the best of any of the brands mentioned and it is also the most expensive.* The

---

* *Consumer Reports,* November 1973, volume 38, number 11, contains a good, objective comparison of subminiature cameras and may be found at the public library.

Figure 4. Minolta 16II subminiature camera in the open position. The Minolta takes a special cassette with 16mm film. This camera has a wide range of shutter speeds and aperture settings.

Kodak 110 Instamatic pocket cameras are not good subminiature cameras from an investigative standpoint because they are considerably larger than the other subminiature cameras and they have a very limited selection of film types from which to choose. While it is possible to obtain subminiature cameras built into such things as cigarette lighters, they are expensive, they do not in many cases take a film size that is easily obtainable, and they can be quite expensive because of the limited market for such items. These items have their place more in fiction than in real life.

One final point to take into consideration when selecting a

Figure 5. Illustrated is a photo that was taken with the Minolta 16II pictured in Figure 4. Plus-X film was used. Note contact print of negative in lower left corner.

subminiature camera: Besides looking for a camera with the features that will suit your needs, consider the system of which it is a part. Does it have accessory filters and lenses? Can one get an adapter for photographing through binoculars? Does the system provide a mini-developing tank, mini-enlarger, mini-tripod, and so forth, thus offering a darkroom and full range of accessories that can be easily transported in an ever-ready case? The camera shown in Figure 4 is a Minolta 16II®. In the closed position it measures $3\frac{1}{8}$ x $1\frac{3}{4}$ x $\frac{15}{16}$ inches. Open it measures $4\frac{1}{4}$ inches in length. The Minolta system offers accessory filters, close-up lenses, a mini-projector, mini-enlarger, flash gun, accessory clamps, etc. Figure 5 shows a photograph taken with the camera pictured in Figure 4. With subminiature cameras, grain and dust will prove to be the biggest problem.

CHAPTER TWO

TELEPHOTO PHOTOGRAPHY

GENERAL CONSIDERATIONS

EXCEPT FOR SOME very distinct problems that are quite characteristic of telephoto photography, much of the art of telephoto photography is simply that of taking pictures. If you have a telephoto lens working at f16, it is very much like any other lens working at f16. Whether you are photographing with an 8mm motion picture camera with a normal lens of about 13mm or a 35mm camera with a standard lens or a 2000mm super-telephoto lens, any specific f-stop is the same in terms of exposure. There is, however, an exception to this rule that one should be aware of. Lenses employing mirrors to shorten the physical length of the lens by folding the optical (light) path into a zigzag pattern will lose about two thirds of a stop of light value because of the mirrors. If a light meter other than an internal meter in the camera is being used, this factor must be taken into consideration when calculating exposure.

The only distinguishing feature of a telephoto picture is a flat, long perspective. *This is a result of the long camera-to-subject distance and has nothing to do with the lens system being used.* If you take a picture from any given point, the perspective will be the same whether the lens is 50mm or 2000mm in focal length. The flattening of the perspective increases with camera-to-subject distance, not with an increase in lens focal length. One should keep this in mind, for someday the defense counsel may attack a telephoto picture being introduced as evidence claiming it shows distortion and consequently should not be allowed in as evidence. Remember, this squeezed-together effect is not distortion. The typical squeezed-together effect of long perspective is very much in evidence in television showings of baseball games.

TELEPHOTO LENSES

In the area of telephoto photography there are basically three types of lenses one will encounter. They are long focus, telephoto and mirror lenses. The long focus and telephoto are both refractive lenses employing no mirrors so the three types of lenses can really be put into two basic groups, refractive and mirror lenses. The simplest of the three types of lenses, and probably the one the reader will encounter the least number of times in telephoto lenses is the direct objective (long focus) lens. (See Figure 6.) All this lens consists of is a direct objective housed in a tube that eliminates extraneous light and positions the lens at the proper distance from the film plane. This lens, with the exception of lack of compactness, is the best for telephoto photography as it offers maximum light transmission and best definition. With the direct objective system the physical length of the lens is the same as the focal length. This can become a matter of concern when getting into the area of extreme telephoto work. The second lens system is a takeoff of the direct objective lens and it is this lens one will encounter the most frequently. It is a true telephoto lens. It has an objective lens in the front of the lens tube just as the long focus lens does, but in addition it has a negative lens in the rear end of the tube that displaces the point of focus further back, thus offering an effective long focal length in a short package. (See Figure 7.) In many cases, lenses of this type have a physical length of one-half to three-quarters the actual focal length, a definite factor in its favor. With the two above-mentioned lens systems, exposure is controlled by the shutter speed of the camera and the position of the diaphragm. The third type of lens the reader is likely to en-

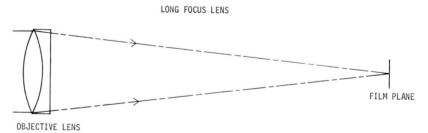

LONG FOCUS LENS

FILM PLANE

OBJECTIVE LENS

Figure 6. Drawing of a long focus lens system. With such a lens, physical length is equal to the focal (optical) length.

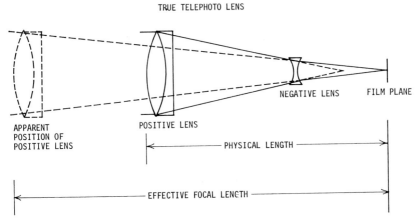

Figure 7. Drawing of a true telephoto lens system. The negative rear element displaces the point of focus further back thus giving the lens an effective long focal length, much longer than the actual physical length of the lens.

counter, and a lens that has some outstanding advantages in the area of surveillance photography, is the reflex-mirror lens. This lens uses a system of mirrors and lenses to fold the optical path three times forming a zigzag pattern. (See Figures 8 and 9.) The

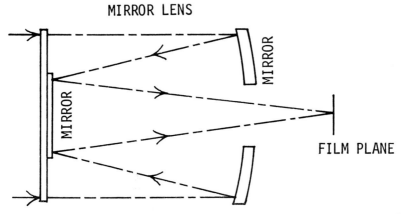

Figure 8. Drawing of a reflex-mirror lens showing the basic optical path folded to provide a long focal length lens in a small package.

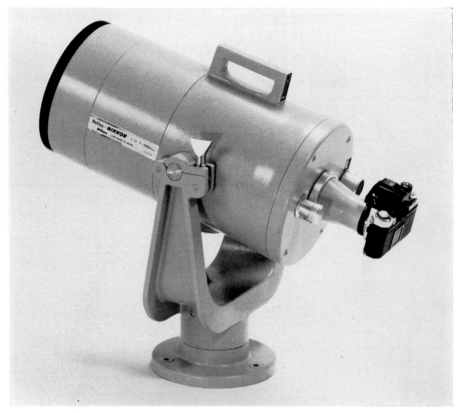

Figure 9. Nikon F2S Photomic® mounted on a 2,000mm f11 Reflex-Nikkor® with mounting yoke. While this lens is 80 inches (2,000mm) long optically, it is only 23⁷⁄₁₆ inches long physically because of its design (Photo courtesy of Ehrenreich Photo-Optical Industries, Inc.).

resulting product is a lens of a long focal length in a very short package. Take for example a lens with a focal length of 500mm. A direct objective lens of this focal length would physically be about twenty inches long, the true telephoto lens (negative rear element) would be between fifteen and seventeen inches long, while the mirror lens would be between 4¾ and 7¼ inches long depending upon the make of the lens. (See Figure 10.) If one is to be taking photographs while hand holding, the mirror lenses are the best choice. Another factor to be considered is the fact that when working with mirror lenses one is not as conspicuous to onlookers as he would be if using a refractive lens of the same focal length. Mirror lenses are also much more maneuverable than the refractive lenses because of their small size.

Unfortunately, mirror lenses are not generally as fast as refractive lenses (the maximum aperture is not as large). In addition to this, there is a further light loss of about two thirds of an f-stop because the mirrors (there are two of them) are not capable of reflecting more than about 85 percent of the light that strikes their surface. Unlike the refractive lenses, mirror lenses do not have a diaphragm to control the amount of light that reaches the film. In many mirror lenses, neutral density filters are built into a turret that is mounted in the rear of the lens which can be rotated to position the desired filter to aid in exposure control. Some manufacturers providing this turret with built in filters will provide filters other than those of neutral density. This tends to be the case when the lens happens to be

Figure 10. Comparison photo of true telephoto lens of 500mm focal length and a reflex-mirror lens of 500mm focal length.

fairly slow, such as Nikon's 1000mm f11 Mirror Lens®. The filters often provided with these lenses are UV haze, yellow, orange and red. Many 500mm mirror lenses which are faster, generally f8, have two neutral density filters to aid in exposure control. Another thing mirror lenses do that refractive lenses do not is cause a vignetting of the image. The vignetting is not a critical problem as its occurrence is not sufficient to be detrimental, and besides, with mirror lenses, so much is gained and so little lost. Mirror lenses also have a very close minimum focus, much more so than is possible with refractive lenses. While a mirror lens with a focal length of 500mm can focus down to about ten feet, a refractive lens of 500mm is only capable of focusing down to about thirty-five feet. This question of minimum focus has little effect however on one engaged in long-range telephoto surveillance.

PRESET vs. AUTOMATIC LENSES

As was stated previously, refractive lenses use a diaphragm to aid in exposure control. Since it is easier to obtain correct focus with the lens diaphragm wide open because the image is then brighter, one finds it is then necessary to stop the diaphragm down the proper amount to coordinate with the shutter speed, so that correct exposure is achieved after focusing. Some lenses do this automatically after the photographer has preselected the f-setting while other lenses leave this to the photographer entirely. There probably would not be too much debate as to which system is more convenient: the automatic, of course. There is, however, a considerable cost difference between the two types of lenses. The preset (manual) lenses are much less expensive. With the preset lenses, it is simply a matter of adjusting to the idea that after focusing the lens must be stopped down unless the photographer for some reason wants the diaphragm wide open.

MAJOR CAMERA BRAND LENSES vs. INDEPENDENT LENS MANUFACTURERS

This is another area in which one can save a considerable amount of money when purchasing telephoto lenses. While each big name camera manufacturer produces an array of lenses for his system which are very good, they are also very expensive. There are independent lens manufacturers that do not make cameras but concentrate on accessories with a strong emphasis on lenses that fit the major 35mm camera brands. They make quite a complete system of lenses (except the normal lens) all the way from wide-angle to telephoto lenses. While these lenses generaly perform as well optically as those put out by the major camera manufacturers, they will in some cases not meet the same specifications concerning mechanical construction. They are, however, very good lenses and, if treated as any fine piece of optical equipment should be treated, will give many years of satisfactory service. The lenses being referred to are such brands as Vivitar®, Soligor®, Camron®, Tamron®, etc. Vivitar lenses are guaranteed for five years from the date of purchase against mechanical and optical defects. Soligor guarantees their lenses for three years. It is worth noting that in several cases the lenses sold by a major camera manufacturer and an independent lens manufacturer are made by the same company and differ only in the name that is printed onto the lens.

COMPOUND REFRACTIVE LENSES

While this type of lens is used to a large degree by astronomers and amateur astronomers, it is generally unknown to photographers. (See Figure 11.) Compound refractive lenses are basically a cross between a direct objective lens and a mirror lens. While a compound refractive lens of say 2000mm focal length cannot be put into quite as small a package as a 2000mm mirror lens, it is only about one-third the length of a direct objective lens of the same focal length. Technically, a compound refractive lens is actually a direct objective (refractive) lens with the optical (light) path divided into three parts by employing the use of two flat mirrors. (See Figure 12.) The light passes through the objective lens in the front and is then intercepted by a mirror after traveling one-third the distance to the film plane. The mirror sends the light back in almost the direction from which it came, almost but not quite straight back, in which case it

Figure 11. 2,000mm f16 compound refractive lens built by the author. The focusing mount is a common bellows extension that was intended for photomacrography. Lenses such as this are relatively easily built if one has any ability at all in woodworking.

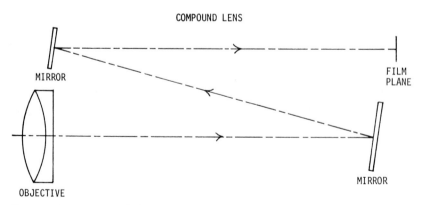

Figure 12. Drawing of the optical path of a compound refractive lens. Note that this lens is simply a long focus lens that has had the optical path folded by two flat mirrors to shorten the physical length.

Figure 13. Subject photographed at one-quarter mile using the 2,000 compound refractive lens shown in Figure 11. Tri-X film was used in conjunction with a number 15 yellow filter.

Figure 14A. Subject vehicle photographed at more than one-half mile using the 2,000mm f16 lens shown in Figure 11. Figure 14B shows the same scene taken with a normal (50mm) lens. Although the time of year is different, both photos were taken from the same camera position.

would send the light right back out through the lens. After the light has been sent towards the front of the lens system by a distance of again one-third the focal length, a second mirror intercepts it, sending it back towards the rear of the lens where it then reaches the film plane.

The greatest advantage of the compound refractive lens is the low cost. Although it is a lens one must build oneself, it is not difficult to make if one has any ability at all in woodworking and cares to take the time. As for the cost, a homemade compound lens of say 2000mm will cost less than half the price of the least expensive mirror or refractive lens of equal focal length on the market and the results it yields will be very good if quality optics and mirrors are used and the construction of the housing is good. Care must be taken when making a compound lens (this holds true for any lens) so that the light baffles are effective in controlling the light path and extraneous light is not permitted to reflect around inside the housing as this will greatly reduce contrast and image sharpness. Super flat black paint or flock paper should be used to line the housing and cover the baffles. Whether or not a diaphragm for exposure control is incorporated into this system is entirely up to the builder. The one illustrated has none. Figure 13 illustrates the kind of results one can expect to obtain with a good 2000mm lens when photographing people at a distance of one-fourth mile. Figure 14 shows a subject vehicle photographed from just over a half mile away. The comparison photograph in Figure 14A, although taken at a different time of year, was taken from the same camera position using a normal lens.

Although it may seem difficult to build one's own super telephoto lens, recognize that it is very common for amateur astronomers to do so with remarkable results. One can consult the public library, as it is full of books on telescope building. They will answer specific questions concerning the actual physical construction of telescopes and telephoto lenses.

TELE-EXTENDERS

It would be advisable and desirable for large departments or agencies which can afford it and which have a strong and consistent need in the area of surveillance photography to purchase a complete line of lenses and equipment. For the smaller departments that do have some need but would encounter difficulty in justifying such an expenditure, there are ways to get around the need of an elaborate array of equipment and still get the job done. As has been previously mentioned, one should buy preset telephoto lenses rather than automatics and also buy lenses other than those put out by the major camera manufacturers as the former are much less expensive. Also, one may wish to consider the use of tele-extenders. A tele-extender is simply an optical device that goes between the camera body and the lens to increase the effective focal length of the prime lens being used. There are four general sizes of tele-extenders. There are extenders that increase the effective focal length by 1.5 times, two times, and three times, and zoom extenders that go from two to three times. Tele-extenders are not produced by the major camera companies as they do cause some image deterioration and none of these companies wants an item associated with their name if it is guaranteed to reduce image quality. Why then, do such companies as Vivitar and Soligor make them? When using a tele-extender to double the focal length of a 500mm lens, for example, the effective focal length is 1000mm and can provide some pretty good results. It will not, however, in all likelihood provide the results one could expect to obtain with a 1000mm prime lens, but good results nonetheless. Another negative factor of tele-extenders is that they decrease the effective aperture by the same amount they increase the effective focal length. In other words, an extender that doubles the effective focal length of a lens also requires two stops extra exposure. A tripler requires three stops extra exposure.

Tele-extenders will also cause a decrease in resolving power of the prime lens with which it is used, as has been previously stated. The loss of resolving power is greater towards the edge of the field than in the center. Resolving power in the center, however, can remain close to 80 or 90 percent of the original resolving power. One way to partially overcome the lack of sharpness around the edges is to stop the aperture down thus obtain-

Figure 15. Subject photographed at one-quarter mile using a Vivitar 800mm f8 lens with a 3X tele-extender. Photographed with Tri-X film rated at E.I. 1,600.

ing better depth of focus. This helps since there is a lack of sharpness around the edges because optically the extender is not capable of focusing the entire image on the exact plane. Optically, that is expecting too much considering the job the extender is supposed to do. It is simply a case of not being able to eliminate spherical aberrations entirely. The problem is greater with a tripler than with a doubler. Another factor concerning tele-extenders that should be mentioned even though it will not generally concern the surveillance photographer is the fact that tele-extenders were designed and intended to be used on the longer telephoto lenses. Because of this, tele-extenders will not work as well on normal lenses and very short telephoto lenses of around 85mm or so. They should start working better on 135mm or 200 mm and longer. A look at some of the more positive factors concerning tele-extenders follows.

When using an extender on a lens, the minimum focus does not change, nor do the distance markings on the lens barrel. One might wonder if he would come out ahead not to use an extender but just to enlarge that much more in the darkroom. This does not hold true because the grain in the film becomes quite noticeable and the image detail simply is not there. The author has experimented, and has also seen the results of others performing the kind of tests whereby a lens with an extender is compared against a prime lens of the same focal length as the effective focal length of the lens with the extender. In almost all cases there is very little difference between the two comparison photos. In many cases it was extremely difficult to detect a loss of quality when the lens with the extender was stopped down for maximum sharpness and five-by-seven-inch prints were made from the resulting negatives. It is sometimes advantageous to use a fast film or to push-process Tri-X film in order to be able to stop down the lens for sharpness. Even though the grain of the film is increased by the pushing, one often gains more than he loses.

It has been claimed by some that while there is an advantage in using a doubler (2X extender), too much is lost when using triplers, and that they are not worth it. The author disagrees

Figure 16A. 50mm lens.

Figure 16B. 135mm lens.

Figure 16C. 400mm lens.

Figure 16D. 500mm lens.

Figure 16E. 800mm lens.

Figure 16F. 2,000mm lens.

Figure 16A through 16F. Figure 16A through 16F shows a comparative magnification of various telephoto lenses coupled to a 35mm single lens reflex camera body. The house on the far side of the lake is just over one-half mile. The photo made with a 50mm (normal) lens is included for comparison purposes.

with this as he has in numerous instances obtained results with a tripler on a prime lens that would not have been possible with a doubler on the same lens. When a 500mm lens with a doubler offers unquestionable subject identification at up to one-eighth mile and the same lens with a tripler offers unquestionable subject identification up to one-fifth mile, and an 800mm lens with a tripler offers unquestionable subject identification as far away as one-fourth mile (the author accomplished this with Vivitar optics; see Fig. 15), who can argue with such results or fault the extenders or lenses being used? Also worthy of mention is the fact that readable license plate numbers have been photographed on vehicles at distances in excess of two miles using a Questar® lens of 1600mm focal length in conjunction with two tele-extenders coupled together. Questar lenses are of extremely high quality but they are also quite expensive. Why cannot the small department or agency without the means to justify the expenditure necessary for an elaborate array of lenses simply supplement its 35mm camera and normal lens with a 200mm and a 500mm preset lens along with a 2X tele-extender? They would then have a decent telephoto capability at a very reasonable cost. Consider also that most of the photographing to be done in urban areas will be at distances of only about 100 to 500 feet.

Figures 16A through 16F illustrate the comparative magnification of various telephoto lenses coupled to a 35mm single lens reflex body. The house on the far side of the lake is somewhat more than a half mile away. All photographs were taken from the same camera position. The photo made with the 50mm (normal) lens is included for comparison purposes.

AIR TURBULENCE AND LIGHT SCATTER

Telephoto lenses of 1000mm or more will cause problems with atmosphere. Looking at the ground glass with a camera focused at some distant object will prove the fact that air is visible and moves to a great degree. The problems caused by turbulence of the air become greater with increased focal length. When you get up to about 2000mm focal length, you have exceeded the maximum for everyday use. Let us examine the problems presented by the atmosphere and how these problems can be overcome to some degree.

As the sun heats the earth's surface, the air near the ground expands, gets lighter, and then rises, being replaced with cooler air from adjoining areas. As a result, the density of the air is not uniform thus causing light rays going from the subject to the camera to refract or bend causing an unsharp and somewhat distorted image. Once realizing and understanding this, there are a number of things one can do to eliminate or at least minimize the problem. While it is true that in surveillance photography the circumstances will often dictate when and where photographs are taken, the photographer should make an effort to take advantage of all factors working in his favor.

Air turbulence is not too great in the early morning hours before the sun has had a chance to heat the earth's surface, so if it is possible, photographing should be considered at that time. If there is more than one prospective vantage point from which to choose, there are a number of factors one should take into consideration when making the selection. First, if exposures are made from a high place such as on a hill or a building, looking down onto the subject, much of the air turbulence that is almost always present close to the ground, even on cool days, can be avoided. Secondly, the amount of air turbulence over a field will be much less than over a parking lot or looking down a roadway. The amount of air turbulence over snow and water is even less than one would find over a field. In long telephoto work, side lighting is preferable to front or back lighting. This is because light scattering is quite severe with back lighting, and front lighting tends to reduce contrast.

If working with color film, the use of a skylight filter is desirable as it helps to reduce the blueness of distant subjects. If working with black and white film, one should try using a yellow filter. A red filter will do the job better than a yellow filter but rather than losing 1½ stops with the yellow filter, the red reduces the light value by three stops. Once the vantage point has been selected and the photographing is about to begin, it is desirable

to use the shortest exposure times possible as this will help minimize the amount of image diffusion from air turbulence. It will
also help control the problems of camera and lens movement. If
working with a black and white negative film, depending upon
how bright or overcast the day is, one may want to consider push-
processing the film to a higher exposure index to allow a shorter
exposure time. (See the section of push-processing films.) The
reader should also be sure to use a lens hood to prevent lens flare,
especially if shooting into or towards the sun. There will be
enough negative factors at work without adding to the problems
by neglecting that simple but important item.

PROBLEMS OF LENS SHAKE, CAMERA SHAKE AND FOCUS

Assuming that the problems of air turbulence and light scatter
have been dealt with effectively, the hardships are not over as

Figure 17. Hand holding telephoto lens. Both elbows should be firm against
the body for support.

Figure 18. Bracing a telephoto lens against a solid object for added support.

there are many problems yet to be solved. They, too, have remedies that can be quite effective. The best optics in the world will not avail if the focus is not accurate and the lens and camera are permitted to shake and vibrate causing image blur. Generally, lenses as large as 500mm can be hand held if a fast film such as Tri-X (400 ASA) is used on a bright day that allows a shutter speed of ⅟₅₀₀ second or more. A mirror lens is much better suited for hand held shots than a refractive lens. In fact a mirror lens as long as 1000mm can be hand held but that is the maximum. A good rule of thumb to remember in hand holding lenses is to use a shutter speed equal to the focal length of the lens being used. That is to say, if one is using a 135mm lens, he should use a shutter speed of ⅟₁₂₅ second. With a 200mm lens, one should use a shutter speed of ⅟₂₅₀ second, and so forth. If this rule of thumb is followed, and if when making the exposure one treats the camera as he would when shooting a rifle (take a deep breath, let part of it out, hold breath and release the shutter gently), a very good percentage of success in obtaining sharp photographs should result. As for the proper method of holding the camera, the lens should be allowed to rest in the palm of the left hand which also focuses the lenses; the body of the camera is secured by the right hand which also takes care of the shutter release and film advance lever. Both elbows should be firm against the body for support. (See Figure 17.) Another technique that is helpful for hand held shots is to brace or rest the camera and lens against some solid object such as a pole, tree or door frame while making an exposure. (See Figure 18.) This is also helpful when working under low light levels that require a slow shutter speed.

When using a lens that is too long to hand hold, some other means must be employed to ensure a stable setup. Naturally a good solid tripod is the most desirable route to take. Also worthy of consideration is employing two tripods. (See Figure 19.) Unfortunately, tripods will sometimes be too obtrusive and cumbersome for the conditions under which one must secure the

Figure 19. Utilizing two tripods is often helpful in extreme telephoto work. A cable release is also desirable.

Figure 20. Belt pod is a useful aid when working with short to moderate telephoto lenses, still or motion pictures.

photographs. By studying Figures 20 through 25 it can be seen that there are several mechanical aids from which to choose. The belt pod shown in Figure 20 is good for short to moderate telephoto lenses, but does not prove effective for long telephoto lenses or for motion pictures employing long lenses of more than moderate focal length. For very long lenses, the stability needed is simply not afforded by the belt pod. When shooting motion pictures, the breathing of the photographer, even though slight, causes the subject to float around in the frame. The longer the lens the more radically the subject will move about. For the motion picture camera a better alternative would be a monopod. The belt pod can be effectively used as a monopod. (See Figure 21.)

When working from an automobile, one might want to consider using a window mount. (See Figure 22.) When using a set-

Figure 21. Belt pod being used as a mono-pod, a technique that often works well with long lenses. Shown is a Vivitar 800mm f8 lens.

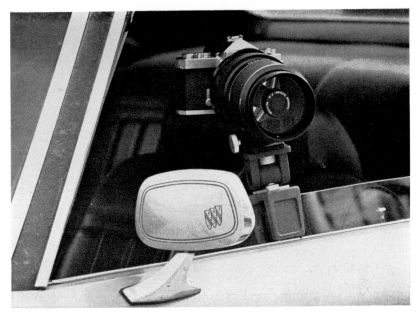

Figure 22. Window mounts are very useful when working from an automobile. Be sure to shut off the engine as its vibrations will have a devastating effect on sharpness.

Figure 23. Gun stock mount is very effective when working with telephoto lenses. Here it is being used with a 500mm f8 reflex-mirror lens.

up such as this, one must be sure to shut the engine off as the vibrations caused by it will in all likeliness make one's efforts a waste of time. As can be seen from the illustration, a vehicle with a dark interior is preferable to a light interior. The camera is not as noticeable. One may also consider using a gun stock mount. (See Figure 23.) When working from an automobile, if there is not a mount of some sort to use, one may try resting the end of the lens on the steering wheel and photographing through the windshield. A variation of this is to rest the end of the lens on the side window and hold the camera body. The window can be raised or lowered to assist in maintaining a comfortable position while having the correct elevation of the lens. This has an advantage over photographing through the windshield, which does reduce light and resolution. When shooting out the side window, one can try sitting in the back seat, a less conspicuous spot than the front seat. (See Figure 24.) When working from an automobile, a beanbag or sandbag can often

Figure 24. Photographer sitting in the back seat and resting the lens on the window for support. Note arrow.

Figure 25. Sandbags and beanbags can prove to be useful to rest a lens on for support.

prove to be an effective aid. (See Figure 25.) If the bag is fixed with a drawstring or zipper, it can be carried empty as sand or gravel can often be obtained on location.

In the area of *extreme telephoto* photography, say anywhere from 500mm to 2000mm focal length, in many cases it will not be possible to work from an automobile and remain at all discreet. In cases such as this it will be necessary to secure a good vantage point within or on top of a building or perhaps to work from a common-looking truck equipped with one-way glass. Also, a sturdy support or mount of some sort will prove to be essential to hold the lens still. When using a tripod, one can try draping a weight such as a sandbag over the lens, or try hanging a gadget bag on the tripod to make fewer vibrations. (See Figures 26 and 27.) The use of a cable release is advisable in order to make an exposure without causing camera shake or movement. Often the mirror in a single lens reflex camera, popping up

Figure 26. Sandbag placed on lens to dampen vibrations.

Figure 27. Weight of gadget bag being used to increase the stability of the tripod against wind, etc.

when an exposure is made, will cause some vibration and a loss of image sharpness. There are two simple things one may consider doing to overcome this problem. First is to make use of the self-timer built into many 35mm SLR cameras. When the timer is set (generally one may select a time from three to ten seconds or thereabouts) and the shutter release button depressed, the mirror pops up so as not to cause an obstruction for the light traveling from the lens to the film. The mirror then stays up until after the exposure has been made. If a time of, say, three seconds is selected, when the shutter release button is depressed, the mirror will pop up. Then three seconds later the shutter will be released to make the exposure. The three-second interval between the mirror popping up and the shutter being released allows for any vibrations caused by the mirror to subside. This has a drawback in that it may be necessary to make an exposure *immediately* in order to obtain the necessary evidence being sought. A second technique may then be in order. With the lens on a tripod, one frames the area to be covered, focuses the lens, then secures the tripod. After this has been accomplished, one locks the mirror in the up position. The mirror, being in this position will not move and cause vibrations when the shutter is released. With the mirror up, one will not be able to observe the subject through the camera viewfinder so it will, in many cases, be necessary to use binoculars. It is also advisable to use a cable release, for without it, one's efforts will probably be defeated. It might not hurt to mount the binoculars on a second tripod with a binocular clamp. There is a second advantage in looking through binoculars. When looking through a camera and lens for several hours one becomes tired and at the end of the day or night his vision will be blurry. It is easier to look for extended periods of time through binoculars than to squint through a camera. Also, the image in the binoculars (7×50) is brighter than what is to been seen through the camera. When working in the area of extreme telephoto photography, it will not be possible to observe with the naked eye what is happening; therefore, the use of optical magnification is essential.

Critical focus is of importance in telephoto photography. If

Figure 28. Eyepiece magnifier provides a 2× magnification of image for critical focusing. It then swings up so that proper framing can be accomplished. A very useful accessory for telephotography and work in low light levels.

one is using a preset lens, it is helpful to open the diaphragm of the lens all the way to afford a brighter image to focus. Also, one will do well to use the ground glass collar around the microprism or split image in the center of the viewfinder, as they do not work with long telephoto lenses. A great aid in critical focusing is an eyepiece magnifier. It is a relatively inexpensive item that screws onto the camera in place of the eyepiece and magnifies the central portion of the image by about two times. It is made to swing up out of the way after focusing so that proper framing can be accomplished. (See Figure 28.)

One last word about telephoto photography. Nothing is to be gained by using a lens of greater focal length than is necessary to fulfill the needs of the specific operation at hand. As has already been stated, the longer the lens, the greater the problems of lens vibration, maneuverability, limited depth of field, and so on.

CHAPTER **THREE**

BLACK AND WHITE FILM
VS. COLOR FILM

WHEN WORKING WITH either still or motion pictures, one must decide whether to work with color or black and white film. This is generally not a difficult decision to make as the situation and the nature of the evidence desired will in most cases dictate which should be used.

If you will be photographing at night using available lighting, you will need an ultrahigh speed film. The most logical choice in this situation would most often be black and white. If color is deemed desirable, one would likely go to High Speed Ekta-chrome® or Kodak Ektachrome EF which is a color slide (rever-sal) film. If you are photographing under normal daylight condi-tions, and colors such as subjects' clothing, vehicles, and so forth, are of importance, you would naturally go to a color film. If, on the other hand, nothing more than good subject identification and/or subject activity is of importance, a good black and white film should prove to be satisfactory as well as more economical. It should be remembered that a good, fine grain, black and white film will yield a sharper image than will color film.

Finally, it cannot be argued that color films do create a more true to life resemblance of a scene than do black and white films. As stated previously, the needs of the particular situation will generally dictate when to use black and white or color film.

CHAPTER FOUR

SURVEILLANCE PHOTOGRAPHY AT NIGHT USING ULTRAHIGH SPEED FILMS

THERE IS NO EASY surefire way to get a picture of a subject at night when using only available lights such as street lights, yard or porch lights, neon lights, the illumination provided briefly by passing automobiles, and so on, nor is exposure calculation anything but difficult in many of the situations you will encounter. There are several good 1°/21° light meters available. A 1°/21° spot meter by Honeywell Pentax is shown in Figure 29. Figure 30 shows the viewing screen with high and low light-scale calibrations, battery checker mark and one-degree center spot. The 1°/21° simply means that while you see a field of twenty-one degrees when looking through the meter, the meter sees and is taking a reading from a one-degree angle of acceptance. The meters are very sensitive to low light levels. Meters of this type are made by several companies such as Honeywell Pentax, Minolta, Soligor, and others. They are, however, quite expensive and will not, in many situations of extremely low light level, eliminate the need for making exposure tests and bracketing your exposures. An area of surveillance photography where these meters are very useful is at night, using available lights when sitting across the street photographing people inside a normally lighted building such as a store. The extremely narrow angle of acceptance and the high sensitivity to low light levels make these meters a valuable tool, allowing the photographer to obtain an exposure reading from a subject across the street. As a general practice, however, since accurate meter readings are so difficult to obtain under many of the extremely low light levels one will often work with, it is advisable to make exposure tests under various lighting conditions to become familiar with ways to ap-

49

Figure 29. 1°/21° spot meter by Honeywell Pentax.

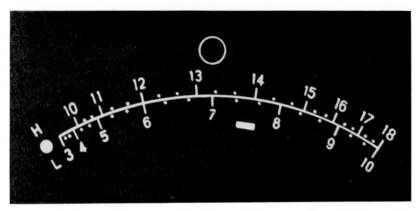

Figure 30. The viewing screen of the Pentax light meter shown in Figure 29. Screen contains high and low light scale calibrations, battery checker mark and one-degree center spot.

proach various situations. It is true that in a situation where photographs will be made of everyone arriving at and leaving a given establishment over a certain period of time, the photographer often has the advantage of going to the location a day ahead of time and making a series of test exposures from the selected vantage point. This, however, generally will not be possible and on-the-spot decisions and estimations will have to be made. One's success or failure in such a situation will hinge very much on how familiar one has made oneself with what to expect from various lighting situations. One can try photographing someone standing under a streetlight (see Figure 31), try sitting across the street and photographing someone in a store, bar or at home. (See Figure 32.) One can experiment with both Kodak Tri-X® film and Kodak 2475 High-Speed Recording film. The reader should take advantage of the information and data supplied in the chapter on push-processing films, experimenting to see what works best for him.

When photographing at night using high speed film and available lights, short to moderate telephoto lenses will generally be the choice. One should remember when using telephoto lenses that they have a tendency to amplify any vibrations or move-

Figure 31. Night photo of subject illuminated by one street light. Tri-X film rated at E.I. 4,000 was used, 135mm f2.8 lens, 1/15 second exposure. Camera-to-subject distance is about 100 feet.

Figure 32. Night photo of subject in normally lit home. Subject was photographed from about 100 feet using a 500mm f6.3 lens. Tri-X film rated at E.I. 4,000 was used.

ments and must be held very still. This problem will be compounded by the fact that the cameraman will be using shutter speeds that are quite slow, probably from ⅟₁₅ second to only as fast as ⅟₃₀ or ⅟₆₀ second. The chapter on telephoto photography deals specifically with the problems of lens and camera shake and ways to deal with these problems. Lenses with focal lengths of 135mm f2.8, 200mm f3.5, 400mm f5.6, and 500mm f6.3 are a realistic variety to consider when dealing with this area of photography. The faster the lens the better. The lens speeds of f5.5 to f6.3 generally will not be effective for a subject standing under a streetlight but will work quite well for subjects in normally lit buildings such as stores. The shorter and faster lenses such as the 135mm f2.8 and the 200mm f3.5 work quite well for subjects under streetlights and also in normally lit buildings. If there is snow on the ground it will act as a reflector of any lights in the area, and photographs will often be much better as a result. Generally the only time a normal lens on the camera might be useful is when one is working from a mobile unit with one-way glass, parked within ten to fifty feet of where the subject will appear, such as in front of a store, bar, and so on. With one-way glass there is generally a light loss of about three f-stops. If using a fast lens of f1.2, f1.4 or f1.8 this should present little or no problem. A normal lens may also be useful if the photographer goes into the building with the subject; however, this would be a unique situation. For work in low light levels, a useful aid for focusing is the eyepiece magnifier discussed in the chapter on telephoto photography. It makes accurate focus much easier. As for what film to choose for this kind of job, unless color is important, black and white film is the better choice as a good black and white film will offer a sharper image and can be push-processed much farther than color reversal films. Color negative films cannot be pushed at all. Refer to the section dealing with push-processing films. You will see that if a very fast color film is needed, High Speed Ektachrome or Ektachrome EF is the best choice.

The following are some basic rules of thumb that may prove

Figure 33A. Greatly enlarged portion of a photo taken at night with illumination provided by one streetlight. Kodak 2475 High Speed Recording Film rated at E.I. 4,000 was used.

Figure 33B. Greatly enlarged portion of a photo taken at the same time, place, and with same equipment as in Figure 33A. Tri-X film rated at E.I. 4,000 was used. Note there is less grain, better subject identification.

to be helpful in low light level photography. They are basically a summary of what has already been stated.

1. Use as slow a shutter speed as conditions will allow.
2. Use the largest aperture possible.
3. If unsure of exposure, it is better to overexpose than to underexpose. Overexposed negatives can be reduced easier and with less loss than would occur in trying to intensify a thin negative. No amount of intensifying will bring out an image that is not on the film. Also, film has greater latitude for overexposure than underexposure.
4. The slow shutter speed may cause problems of unsharpness. Good support is essential.
5. Long focal length contributes to blur by magnifying movement. Again, good support is essential.
6. Large lens aperture reduces depth of field.
7. Limited depth of field makes accurate focusing critical.
8. Low light level makes focusing very difficult. Use an eyepiece magnifier.
9. IMPORTANT: When photographing a subject that is dimly illuminated by a low light level source, *be mindful that the light source is at least a little bit in front of the subject* so as to avoid back lighting that could result in a silhouette that will have little or no value for identification purposes.
10. When photographing people at night, it is necessary (in most cases) to use a film and processing formula that enable the photographer to give the film an exposure index anywhere from 1600 ASA to 6400 ASA. Figures 33A and 33B show a comparison of two photographs taken under the same conditions at the same time. Figure 33A was taken with Kodak 2475 High Speed Recording film. Figure 33B was taken using Kodak Tri-X negative film. Both films were rated at E.I.4000. Both films were then developed according to the development instructions provided in the chapter on push-processing films.

CHAPTER FIVE

PUSH-PROCESSING PHOTOGRAPHIC FILM

A S MENTIONED PREVIOUSLY, one of the advantages of the 35mm single lens reflex camera as a tool for the surveillance photographer is the fact that there is a wide variety of fast lenses available for these cameras. Since the surveillance photographer must in most cases work with whatever available light exists, he must have available to him in addition to fast lenses, fast films. Often however, one will be faced with conditions where the light level is so low that obtaining a correct exposure with a fast lens and a fast film such as Kodak Tri-X, normally rated, is impossible if a realistic shutter speed is to be used. Since using a flash or strobe unit is out of the question in surreptitious photography, does one then simply put the camera away and try some other time? No! It is in many cases possible to obtain results by push-processing film. This means to pretend the film is of an ASA rating higher than it actually is. For example, while Tri-X film is normally rated at 400 ASA, the photographer feeling that the film is too slow to allow a fast enough shutter speed for the particular job may choose to pretend the film is really an 800 ASA film and expose accordingly. The negative naturally will then be underexposed by one f-stop, so the photographer when developing the film will simply overdevelop to compensate for that underexposure. The degree of overdevelopment must increase the more the film is underexposed. This is accomplished by either increasing the development time, the development temperature, or both.

Unfortunately, push-processing film is a game of compromise, of give-and-take. While it is very much to the photographer's advantage to know how to push-process films, he must realize that grain and contrast increase with pushing. The photographer must be able to discern when to push his film and how far to push it. While it is true that by pushing one can often obtain a photograph of much higher quality than would be possible with the film rated at its normal ASA, one will begin to lose what he stands to gain if he pushes farther than is necessary.

Another point worth noting is the fact that when push-processing black and white negative film, there is generally an *actual* increase in film speed of only about one third of an f-stop. When a black and white film is made, it is given an ASA rating. This rating is figured to give the film the minimum amount of exposure necessary to obtain an excellent print of a scene with average contrast. There is, however, a safety margin of about one third of an f-stop figured into the ASA rating of the particular film. Push-processing is possible because overdevelopment will serve to bring out an image that has been underexposed but is on the film nonetheless. Shadowy areas that did not register on the film cannot be brought out by any amount of overdevelopment. The bright portions of a scene naturally can undergo a considerably greater degree of underexposure and still register on the film than can the very dark portions of the scene. Because of the one-third f-stop safety margin of the film, very dark shadowy areas of the scene can suffer only about one-third f-stop of underexposure and still be brought out by overdevelopment. An image that is simply not there cannot be brought out by any amount of overdevelopment. This is why an *actual* film speed increase of only about one-third f-stop is possible. In a surveillance photograph where shadowy details are of minimum importance and subject identification (subject's face) is of prime interest, a great deal of underexposure and overdevelopment can be tolerated.

Now that it has been established that push-processing black and white negative film does not significantly increase the actual film speed or its sensitivity to light, there are also other things a person should be aware of when making the decision to push or not to push. As already stated, when push-processing film, the more it is pushed the greater will be the increase in the contrast and grain of the film. Because of this, one should never push the film any more than is necessary to obtain the results needed. Generally it would be argued that when pushing film, because contrast is increased, the resulting negative should be printed on a soft (low contrast) grade of paper. Normally this would be correct, but one will soon become aware that the rules often applying to general photography will not always apply to surveillance photography. This is true because of the unique and uncommon conditions under which surveillance photography is performed and because the needs or requirements of the photographer are different from those of the general photographer. Going back then to the statement about pushed film being printed on a paper of a low contrast grade, one can experiment for oneself, but will probably find that the best subject identification is achieved when a greatly pushed negative is printed onto a paper with a high contrast grade such as a number four or five. This tends to separate the features of the subject's face, making them more distinct. The value of a surveillance photograph lies in its ability to identify a subject or subjects and to prove or disprove something. Therefore, the criteria by which general photographs are judged often do not apply when it comes to judging surveillance photographs. Keep this in mind when push-processing films; the photographer is after evidence, not a piece of art.

While push-processing does not significantly increase the film speed of black and white films, push-processing color reversal (slide) film does cause an actual increase in the film speed or ASA rating. Reversal films differ from negative films in that negative films get darker with exposure while reversal films get lighter with exposure. Unexposed portions of negative films are light while the unexposed portions or reversal films are very dense or black. A properly exposed reversal film does not use the full potential of the density of the film because if it did, it would end up being too dark to be viewed on a screen after a normal development process. As a result, reversal films have a lot of density potential which is not taken advantage of and not needed in normally exposed and processed films, even though the film is quite capable of recording details at much higher densities. In short, *a reversal film that has been underexposed and as a result is too dark for viewing often has recorded and contains both shadow and high light detail.*

By increasing the first development time, the exposed areas of the film can be lightened and the film is then light enough to be viewed normally on a screen when projected with a slide projector. By increasing the first development time the proper amount and thus making the film lighter, a full compensation for underexposure has been achieved and an increase in the film speed is the result. Remember when push-processing film, only the first development time is increased. There are two development stages in the processing of color reversal films. Also, when pushing reversal film, there is a loss of image quality just as there is when pushing black and white negative films. There will be an increase in grain, the maximum density will be reduced and there will usually be a slight change in color due to the fact that all the properties of the film's emulsion do not react the same way to the increased development.

If a very fast color film is needed for a particular job, one should consider using High Speed Ektachrome by Kodak. The ASA rating of High Speed Ektachrome Tungsten® is 125 ASA, daylight is 160 ASA. If a higher ASA is needed, one should use either one of the High Speed Ektachrome films, as commercial labs (many of them) will push the Tungsten to 320 ASA and the daylight film to 400 ASA. If one wishes to purchase an E-4 developing kit from a photographic outlet and process the film himself, he can go much farther and has the option of taking advantage of Kodak Ektachrome-X® film 64 ASA in addition to the High Speed Ektachrome films. The Ektachrome-X film as well as the two High Speed Ektachrome films can be pushed much faster if one purchases an E-4 kit and does the processing himself rather than using a commercial lab service. (See the development times furnished in this chapter.) It is not possible to alter the film speed by altering the development times on any other Kodak color films besides the three that have been discussed and the Ektachrome EF. If one needs a very fast color film for prints, the best choice will likely be to use one of the reversal films that have been discussed, and then photograph the resulting slide. When doing so, however, there will be a further deviation in color and the print will usually be somewhat darker than the slide from which it was taken.

Table I shows a list of developing formulas for push-processing various high-speed films. One should remember to go very gently on the degree of agitation as agitation increases grain and contrast. Also, one should remember not to push a film any fur-

TABLE I

FORMULAS FOR PUSH-PROCESSING PHOTOGRAPHIC FILMS

Kodax Tri-X Film. Agitate at 60-second intervals.
  400 ASA Normal development.
E.I. 800 ASA D-76, 12 min. at 68°f.
E.I. 1200 ASA D-76, 9¾ min. at 75°f. *or* HC-110 (dil. A), 6½ min. at 68°f. *or* Acufine® developer, 5¼ min. at 68°f.
E.I. 1600 ASA D-76, 13 min. at 75°f *or* HC-110 (dil. A), 8 min. at 68°f.
E.I. 2400 ASA Diafine®, two step development process, see instructions on package.
E.I. 4000 ASA HC-110 replenisher, 1:15 (one part replenisher to 15 parts water). Develop for 8 min at 75°f. Agitate at 3 and 6 min. Discard at 8 min. EXPERIMENT A BIT AS YOU MAY WANT TO ALTER THE LENGTH OF DEVELOPMENT TIME AND THE AGITATION INTERVALS TO SUIT YOU BEST.
Kodak 2475 High-Speed Recording Film.
E.I. 4000 ASA DK-50, 9 min. at 68°f.
E.I. 6400 ASA Diafine, mix solutions A & B at 70°f.
      Develop in solution A for 3 min. and then in solution B for 3 min., rinse well in running water for 1 min. at about 70°f and redevelop 2 min. in both solutions A & B respectively. Rinse, fix and dry.

Kodak High Speed Ektachrome Film. Follow instructions provided with E-4 Chemicals.

Extend first development times only.	(Daylight)	(Tungsten)
Normal Development	160 ASA	125 ASA
Normal plus 35% increase	E.I. 320 ASA	E.I. 250 ASA
Normal plus 75% increase	E.I. 640 ASA	E.I. 500 ASA

Kodak Ektachrome-X Film. Follow instructions provided with E-4 chemicals. Extend first development time only.

Normal Development	64 ASA
Normal plus 35% increase	E.I. 125 ASA
Normal plus 75% increase	E.I. 250 ASA

Note: The above formulas are general guides and may be altered some to better suit individual needs and techniques. This holds true especially for Tri-X rated at E.I. 4000 ASA.

ther than is necessary to obtain the results desired, since the farther a film is pushed the more one loses to grain, contrast, and so forth.

Reciprocity-effect adjustments have not been discussed here because they concern a person only when making very long exposures of about a second or more, or exposures of much shorter than one thousandth of a second.*

---

* For information on this topic, the reader should order Kodak Publication Number 0-2 (Reciprocity Data, Kodak Professional Black and White Films).

CHAPTER SIX

INFRARED SURVEILLANCE PHOTOGRAPHY

THE LIGHT SPECTRUM

THE VISIBLE LIGHT spectrum is made up of various wavelengths of electromagnetic radiation. The spectrum is made up of violet light on one end with a wavelength of about 400 millimicrons. As the wavelengths get longer we get into blue, then green, yellow, orange and finally deep red which is about 700 millimicrons. Beyond the two extremes at each end of the spectrum is electromagnetic radiation which continues to get shorter in wavelength on the violet end and longer on the red end. Infrared photography takes place in the region just beyond the red end of the spectrum between about 700 and 900 millimicrons. This region is not visible to the human eye. While the spectrum does continue far beyond 900 millimicrons, it has nothing to do with infrared surveillance photography and will not be discussed here.

BASIC TECHNIQUES AND EQUIPMENT, 35mm STILLS

Basically all one needs to surreptitiously take an infrared photo of someone in the dark is any 35mm camera, a roll of Kodak High Speed Infrared® film, a gelatin Kodak Wratten Filter Number 87®, and an electronic strobe unit. The higher the BCPS (candlepower) rating of the strobe the better. The film is a special film that is sensitive to a range of electromagnetic radiation between 700 and 900 millimicrons. Unfortunately this film is also sensitive to the visible region of the spectrum; this is why the filter is needed. Also, the film cassette cannot be taken out of the canister except in total darkness; consequently the camera must be loaded and unloaded in total darkness. It is advisable to take a changing bag into the field for this purpose. The infrared filter allows electromagnetic radiation just beyond

700 millimicrons to pass through but stops the visible light which is less than 700 millimicrons. When a filter-covered strobe is flashed, if one is in the dark looking directly at it he will see a very faint, very brief, dull red glow. There are many fields besides the investigative field that use infrared photography as a scientific tool under daytime conditions. They often put the filter over the lens of the camera to prevent visible light from reaching the film when an exposure is made. The filter does, however, allow the infrared radiation to pass through to the film.

In surveillance photography, however, it is done a little differently. Because it is dark when one is using infrared film, he does not have to worry about visible light interfering with the film when an exposure is made without a filter over the lens. A streetlight or porch light a short distance away will not cause any problems. The filter over the strobe unit is necessary to filter out the visible light; however, if one did not have it, it takes little imagination to guess just how the subject would react immediately following the first exposure made from a carefully selected vantage point. As previously stated, the filter can be obtained in gelatin form. Such a filter can easily be cut to size and taped to the face of the strobe unit. Because gelatin filters are difficult to clean, it is desirable to mount a small piece of glass over the filter to protect it. When dirty, the glass can be easily cleaned.

When focusing for an infrared photograph, one must make an adjustment or a sharp image will not be possible since the infrared rays are longer than the visible light rays, thus focusing at a different point. Most lenses made for 35mm single lens reflex cameras have an infrared, auxiliary focusing mark that allows one to easily make this correction. If a lens has this, after focusing with visible light, one should take the part of the focusing ring that lines up with the infinity mark and rotate the ring until that mark lines up with the infrared focusing mark. No further adjustment should be necessary. If conditions are such that one can stop down, thus having more depth of field, this is to one's advantage. If there is no auxiliary focusing mark, one should make tests. Generally, to get a correct focus with infrared radiation, extending the point of focus (film plane) by one fourth of one percent of the focal length of the lens will result in a sharp image. This is what happens when one uses the auxiliary focusing mark on the lens. If it is known ahead of time what the camera-to-subject distance is going to be, one should just focus on something at that distance, make the correction and tape the focusing ring in position so it does not move. If it is not known what the camera-to-subject distance will be, when the subject appears, the photographer should try focusing on something about as far from him as the subject. He should focus on such things as a streetlight, a porch light, or anything that it is possible to focus upon. Finally, he should adjust the focus and shoot. If there is nothing to focus upon, one has no alternative but to estimate the distance and hope for the best. It will also be necessary to establish a guide number for the strobe and filter assembly. To establish a guide number, just take photographs of an object at any known distance (20 feet for example) using all the f-settings on the lens. After processing the film, determine which f-setting gave the best exposure and multiply that number by the camera-to-subject distance. For example, if the object photographed was twenty feet away and the best exposure was achieved at f-4, the guide number is eighty ($20 \times 4 = 80$). Any time thereafter that one wishes to make an infrared exposure using that strobe and filter, he should just divide the camera-to-subject distance into eighty and the answer will be the proper f-setting to use. The techniques and equipment that have been discussed are those most commonly used by investigative agencies in the United States. A maximum camera-to-subject distance of about fifty feet will be possible if the above instructions are followed. It is possible to purchase special infrared flash bulbs but they offer a range of only about twenty to twenty-five feet.

ADVANCED TECHNIQUES, 35 mm STILLS

For those who feel that a maximum range of fifty feet is just not sufficient to do the job, there is a very inexpensive way to increase this range by three times or more. A distance of 150 feet or more will certainly enable one to do some good. When working at 150 feet or more, a telephoto lens of 400 mm to 500 mm will be essential to obtain an image size large enough to have subject identification. Refractive lenses are generally faster than most mirror lenses of the same focal length. While there are available on the market some very fast mirror lenses, they are generally very expensive. Should the photographer, however, have one of these available to him, he might consider using it, as there is no focusing correction or adjustment necessary with mirror lenses when used for infrared photography.

Next, after the telephoto lens, one will need a telephoto strobe unit that throws a narrow, highly concentrated beam of light as opposed to a wide angle of coverage. The photographer needs all the intensity of radiation he can get. Such devices are available on the market, but they too, just like so many other special-

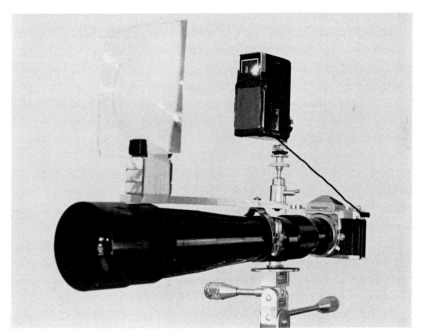

Figure 34. Illustrated is a typical setup for taking telephoto infrared photographs. The radiation source consists of a Fresnel lens mounted in front of a strobe unit covered with an infrared filter to block visible light.

ized pieces of equipment, are very expensive. For about ten to fifteen dollars a normal strobe unit can be made to do this job.*
This lens is mounted 8½ inches in front of the strobe unit by using a binocular clamp (tripod mount) or any other means one may wish to devise. (See Figure 34.) The rough side faces the subject. This lens takes the radiation from the strobe that would normally cover a large area and concentrates it into a strong beam, thus greatly increasing the intensity and consequently causing an increase in the effective camera-to-subject distance. By referring to Figure 35 it can be seen that hot spots will generally result. If these hot spots do not appear in the area that will be covered by the telephoto lens there is no problem. If they do, one should put a diffusion glass of some sort over the front of the strobe unit along with the infrared filter. Some loss in intensity will result, but that cannot be helped. The setup in Figure 34 has no diffusion glass because the hot spots were not causing a problem. One should establish the guide number by using the same method that has already been explained. Once familiar with this setup, the photographer should establish another guide number to be used when the infrared film is going to be push-processed by increasing the development time by 50 percent. With the setup shown in Figure 34, a guide number of 900 with normal process gives an effective camera-to-subject distance of 145 feet. (See Figure 36.) By increasing the development time 50 percent, a guide number of 1145 and a range of 182 feet is possible. (See Figure 37.) As is the case with ultrahigh speed black and white negative films, one will in many cases find himself printing infrared negatives on a high contrast grade of paper to obtain the best subject identification.

Finally, if the photographer is in a pinch because he is forced to work at too great a distance for his equipment to handle, there is yet one last maneuver he can try that may put him within range. By removing the Kodak Wratten Filter Number 87 and replacing it with Kodak Wratten Filter Number 88A®, the setup

---

* A Fresnel® lens with a focal length of 8½ inches (stock no. 40,803; cost, $7.25) can be ordered from Edmund Scientific Company, 620 Edscorp Building, Barrington, New Jersey 08007.

Figure 35. In this photo, the pattern of the infrared illumination provided by the equipment shown in Figure 34 can be seen. Note the hot spots under the subject.

Figure 36. Subject photographed at 145 feet in total visual darkness using equipment shown in Figure 34.

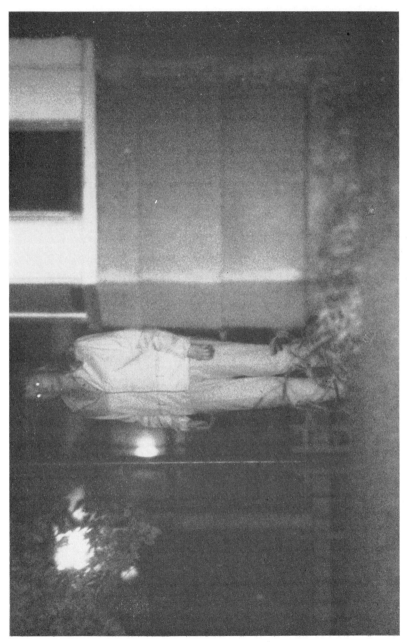

Figure 37. Subject photographed at 182 feet in total visual darkness using the equipment shown in Figure 34. Note porch light about 120 feet beyond the subject that was used as a reference point.

shown in Figure 34 had an increase in camera-to-subject distance from 182 feet to 265 feet with a 50 percent increase in development time. An 88A filter is not recommended by Kodak for surreptitious photography at night, but the author sees no reason why it cannot be used effectively for this kind of work in view of the great distance at which it will be used. The 87 filter cuts the electromagnetic radiation off at about 740 millimicrons; the 88A filter cuts it off at about 735 millimicrons. It was stated earlier that we are capable of seeing only up to about 700 millimicrons, but because the cut of these filters is not really sharp, a weak red glow is detectable. It is true that the amount of glow given off when the 88A filter is used is greater than when the 87 filter is used. However, the difference is not extreme and the increase in the distance at which it will be used makes up for that difference. Unless the subject looks directly at it, he will not detect it. When Kodak made the filter recommendations for surreptitious infrared photography at night, they were probably thinking in terms of a camera-to-subject distance of only about fifty feet or so, not a distance of or in excess of 200 feet. If they did, it is very probable they too would have included in their recommendations the 88A filter.

One of the big problems encountered in infrared surveillance photography is that of not knowing, because of the darkness, when to make the crucial exposure that will record whatever subject action is of interest. If it is known just where the subject will be (perhaps he always parks in the same spot or meets someone in a certain doorway) the camera and lens can be set up on a tripod to cover that area and the subject can be observed with one of the many World War II sniperscopes that are now surplus items because of the invention and implementation of the more sophisticated starlight scope. By watching the subject and depressing the cable release at the right time, one can select one's own exposures. (See Figure 38). Another possibility of the sniperscope is to have a mount for both the camera and the scope. With such a setup, the camera and sniperscope would move together and what the photographer would see through the scope, the film would see through the camera lens. This will not, however, alleviate the problem of focus. One should also re-

Figure 38. Photographer using surplus sniperscope to observe a subject and select exposures after setting up infrared equipment to cover a desired area.

member to consider parallax. If a sniperscope is not available, one will have to guess when the exposures should be made and rely to a great extent on luck. If it is so dark that the subject cannot be seen through the viewfinder, if it is known about where he is in relation to some distant light, and so on, the photographer should use that light as a guide or reference point. The photograph in Figure 37 was done in this manner. Note the porch light on the house about 120 feet beyond the subject.

HONEYWELL PENTAX NOCTA FOR INFRARED PHOTOGRAPHY

In the early 1960's Honeywell Pentax developed a special infrared single lens reflex camera system they call the Honeywell Pentax Nocta®. A front and rear view of the Nocta is shown in Figures 39A and 39B. This camera is equipped with a non-changeable 300mm f3.3 telephoto lens, an image converter sys-

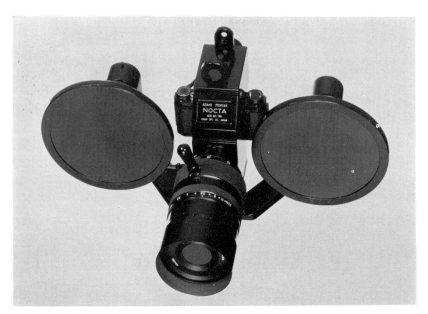

Figure 39A. Front view of the Honeywell Pentax Nocta. This infrared camera is equipped with a 300mm f3.3 lens, an image converter tube, an instantaneous radiation source for film exposures and a continuous radiation source for subject viewing (Photo courtesy of Honeywell Pentax).

Figure 39B. Back view of the Honeywell Pentax Nocta (Photo courtesy of Honeywell Pentax).

tem requiring high voltage that operates on the same principal as the infrared snooperscope, and an instantaneous radiation source is used to expose the film while the continuous source is for observing and framing the subject. The Nocta uses a 35mm format.

With the Nocta it is possible for the photographer to view the subject and select exposures. While the system is capable of photographing up to 300 feet, it is questionable what kind of subject identification one can obtain at that distance with the image size afforded with a 300mm lens.

The Nocta is a system that had real potential as a surveillance tool but sales were hurt by the implementation of the starlight scope. Unlike the Nocta, starlight scopes offer the capability of interchanging lenses. Also, because starlight scopes operate on the principal of light intensification rather than infrared radiation, they do not require the use of an infrared radiation source which in the case of the Nocta happen to be infrared flash bulbs which must be changed after each exposure is made. The continuous source on the Nocta is for observation purposes only as it is too weak for exposing the film sufficiently. Finally, the Nocta was selling for about $5,900 while starlight scopes start at less than $1,500 and go up in price from there.

It was previously stated that infrared flash bulbs provide a camera-to-subject distance of only about twenty to twenty-five feet. The reason the infrared flash bulbs used with the Nocta are capable of such a long range is because a parabolic reflector is used to concentrate the radiation into a strong beam in much the same manner as does the Fresnel lens that was previously discussed.

The Nocta is a good system, but when it is compared with the more sophisticated and modern starlight scopes, feature for feature, and also in price, the starlight scopes become the obvious choice.

IS PHOTOGRAPHING WITH INFRARED WORTH THE TROUBLE?

Photographing with infrared along with a telephoto lens and Fresnel lens does result in a setup that is rather bulky. One should consider, however, that working with telephoto lenses in general, with the exception of 500mm and 1000mm mirror lenses, means using equipment that is bulky and less than convenient. This is also true when using a telephoto lens with a starlight scope. In considering whether or not the effort is worth the bother, the photographer simply must ask himself whether or not the evidence that could be obtained is worth the trouble. Generally it would seem that if the case is important enough to warrant surveillance, the trouble should prove to be inconsequential. Finally, bulkiness really should not prove to be too much trouble in cases where a vantage point is selected, the equipment is set up, and then one must sit down and wait.

CHAPTER SEVEN
INFRARED SURVEILLANCE PHOTOGRAPHY, 16mm MOTION PICTURES

IN MANY RESPECTS making infrared motion pictures is much the same as making infrared stills. The most notable difference is that one is forced to deal with a continuous radiation source as opposed to the instantaneous radiation source that was discussed for still photographs. It was stated that when looking directly at a strobe filtered with a Kodak Wratten Filter Number 87 one could detect a brief, dull red glow when the strobe was flashed in the dark. The big reason that the telltale glow was fairly difficult to detect was because it was of such short duration, only about $\frac{1}{1000}$ second. With a continuous radiation source, however, the dull red glow becomes quite noticeable. Kodak literature states that if a Kodak Wratten Filter Number 87C® is used in place of the 87 filter, this red glow can be eliminated. It is, they claim, eliminated at the expense of about 2½ f-stops loss in exposure, thus reducing the effective camera-to-subject distance by a considerable degree. The author found that while the 87C filter did reduce the intensity of the red glow a considerable degree, it did not eliminate it entirely. It is unlikely, however, that the subject would realize he is being photographed with infrared materials if he did notice the red glow as few people have any real knowledge or awareness of what infrared photography is all about. The danger lies not so much in that the subject will automatically realize he is being photographed, but in that he may wander over to satisfy his curiosity concerning the red glow. The locations and conditions under which one will engage in infrared motion picture photography will influence to a great extent how noticeable the radiation source will be.

One should be mindful of ways to disguise the radiation

77

source when he is examining the area where he will attempt this task. It will in some cases be possible to disguise the radiation source as an EXIT sign or put it where the taillight of a car should be. The glow is not by any means as bright as an automobile taillight but a glow there, if set up right, would cause no suspicion. A person will do well to exploit his imagination to the maximum. If the radiation source is to be mounted with the camera, a spotlight of some sort will be the best choice. The author tried many spotlights and finally settled on a thirteen-volt General Electric Aircraft Landing Light Number 4537® and used it off a twelve-volt car system. That light has a candlepower rating (BCPS) of 200,000. Using it on a twelve-volt system rather than a thirteen-volt system probably reduces the candlepower some. When using this light with a Kodak Wratten Filter Number 87, a guide number of 400 was established giving a maximum range of 142 feet when using an f2.8 lens and an exposure time of ¼₄ second. If one decides to do some experimenting with continuous radiation sources, he should keep in mind the fact that, as a general rule, low voltage bulbs tend to have a higher infrared output than high voltage bulbs. Also, the more directional and concentrated the beam of a spotlight the harder it will be to ensure that the radiation is on the subject. This is the main reason the author chose the bulb mentioned over other spotlights; it covers a greater area yet is strong enough to give an acceptable guide number and camera-to-subject range. When establishing a guide number for the instantaneous radiation source, shutter speed was important only to the extent that it was set so that proper synchronization was ensured. When determining a guide number for a continuous radiation source, the shutter speed is important. Whatever shutter speed is used when the test exposures are made must always be used with the resulting guide number or improper exposures will result.

The problem of subject identification is a very real problem when making infrared movies with the 16mm motion picture camera, since the area of the frame is less than one-ninth that of a 35mm frame. The frame size of a 16mm motion picture camera is about 10mm × 7½mm while the frame size of a 35mm camera is about 36mm × 24mm. As can be seen, the image size will be very small if the lens used is short enough to give an acceptable degree of coverage. If using a lens with enough focal length to give an image size sufficient to secure good subject identification, the image size of the subject would be so large in comparison to the frame size that it would be very difficult to keep the subject properly framed. If the nature of the case warrants it, a possible solution would be to use two photographers, one to use a motion picture camera to document subject activity while the other uses a 35mm camera to record subject identification. If the activity is such that the subject will be in one spot for the period of time he is to be photographed (drug pusher meeting a user consistently in a certain doorway), it would be possible for one photographer working from a good vantage point to handle both the motion picture and the still camera. By having both cameras mounted on tripods and ready to go when the action starts, the motion picture camera can be activated and left to run itself while the photographer begins taking still shots. These are ideas or points to keep in mind as each situation will be unique and one's degree of success will in many respects depend on how imaginative and original he can be.*

---

* For additional information concerning infrared photography one may wish to obtain a copy of Kodak Pamphlet No. M-8 (Criminal Detection Devices Employing Photography) or Kodak Technical Publication M-28 (Applied infrared Photography).

CHAPTER EIGHT

PHOTOGRAPHY AT NIGHT USING STARLIGHT SCOPES

STARLIGHT SCOPES AND HOW THEY DIFFER FROM INFRARED SNIPERSCOPES

W̶HEN CONSIDERING NIGHT vision devices (NVD's), there are two basic types of equipment from which to choose. They are infrared scopes and starlight scopes. Starlight scopes, while having a lot in common with the famous World War II Sniperscopes, are much more advanced and sophisticated. Figure 40 shows a diagram of a sniperscope system and Figure 41 shows a photograph of a sniperscope. Note that the scope utilizes a light source with an infrared filter over it to illuminate the subject with invisible infrared radiation. The electron tube then converts these invisible infrared rays to a visible image after the objective lens has focused them onto the face of the tube. The starlight scope on the other hand does not rely on infrared radia-

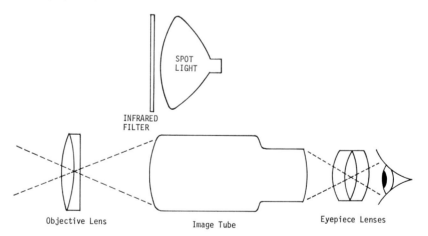

Figure 40. Simple drawing of an infrared sniperscope. Note that a light source and infrared filter are necessary to illuminate the subject. The electron tube then converts the infrared image to a visible image.

Figure 41. Photograph of the famous WWII Sniperscope.

tion, but rather takes the photons that are provided by the stars, streetlights, and so forth, and increases their energy, thus relying on light intensification. Depending upon the make of the scope, the usual light energy increase or gain is between 35,000 and something over 65,000 times. In basic simple terms this is done by changing the light energy to electrical energy, amplifying it, and then changing it back to light energy. In Figure 42 is a drawing of a basic starlight scope. Figure 43 shows three Star-Tron® models.

Starlight scopes are free from several of the drawbacks that are characteristic of the infrared scopes. With the starlight scopes, unlike infrared scopes, a heavy battery pack for the radiation source is not necessary, thus making it much lighter and easier to carry and manipulate. Also the field of view is better with starlight scopes because the operator is not limited to the small field of view and range provided by the spotlights used in infrared radiation sources. Finally, with starlight scopes, one can

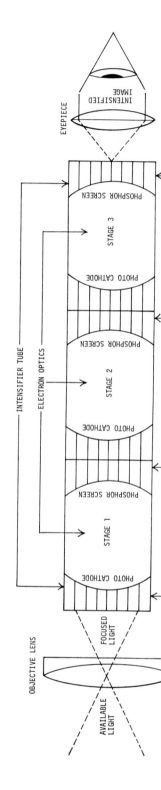

Figure 42. Simple drawing of a starlight scope (first generation). The starlight scopes do not rely on infrared radiation, but rather, operate on the principle of light intensification.

Figure 43. Pictured are Star-Tron scopes models MK 303-A,®, MK 202-A and their pocket scope MK-1®. The Star-Tron system is not, however, limited to the three scopes pictured here (Photo courtesy of Smith & Wesson).

use lenses with a wide variety of focal lengths to provide the image size desired at various distances from a subject. Being able to do this is something that the user of sniperscopes does not enjoy. The only real point in favor of infrared sniperscopes over starlight scopes as night vision devices is they can be purchased in most cases for only a few hundred dollars while the starlight scopes sell from fifteen hundred to several thousand dollars. If one's needs are simply to make visual observations to detect and watch a subject and funds are limited, a sniperscope would be a good investment. If, however, one's needs call for any photographing, a sniperscope is not suitable for this and unless you can photograph with infrared film and use the sniperscope simply to aid in the selection of exposures, you had better consider the starlight scope. While it is true that there are refined versions of the surplus sniperscopes available, they are not suitable for photography and you will also pay nearly as much if not more for them as you would pay for many models of the starlight scopes.

18mm, 25mm AND 40mm IMAGE INTENSIFIER TUBES

There are three standard sizes of starlight scopes available on the civilian market. They are 18mm, 25mm and 40mm sizes. These sizes have to do with the format of the image tubes being used. The 40mm tubes offer the largest image format and consequently they are for that reason the better choice for photographic applications. The 40mm scopes, however, are physically larger and heavier than the 18mm or 25mm scopes, and if size is of great concern, one would not do wrong to invest in the 18mm scopes as they are also very suitable for photographic applications. Figure 44 shows a photograph taken with the Javelin® starlight scope which utilizes an 18mm image tube.

FIRST GENERATION AND SECOND GENERATION IMAGE INTENSIFIER TUBES

If one is considering the purchase of a starlight scope system, it would be wise to secure promotional literature from the various manufacturers or distributors and carefully study the materials. When so doing, one will notice what are referred to as "first generation and second generation" intensifier tubes.

The first generation systems, so called because they were the first to be designed and put into production, have the benefit of a higher degree of resolution than do the second generation units, the resolution of the former being about forty lines per millimeter. A fault of the first generation units is a notable degree of distortion around the edge of the field of view.

The second generation systems are considerably smaller and more compact than are the first generation systems and suffer only about one-fifth the distortion that is characteristic of the first generation systems. The resolution, however, is not as good

Figure 44. Illustrated is a photo taken through a Javelin Model 221® (First Generation) Night Viewer by a California Police Department through the windshield of a squad car. Exposure was 1/60 second on Tri-X film rated at E.I. 1,200 and developed in Acufine. Nearest streetlight was half a block away (seen in photo upper left). Time of day was about one A.M. The lens was set at F4 (Photo courtesy of Javelin, Division of Apollo Lasers).

with the second generation systems. Finally, one will pay more for the second generation systems as the cost of manufacturing the image intensifier tubes is considerably higher than with the former.

When deciding upon a first or second generation system, one must do just as he had to do when deciding whether or not he needed an 18mm, 25mm or 40mm image intensifier tube. The main question is whether or not the unit will be used exclusively for photographic applications. If the unit is intended to serve a wide range of needs, then ultimate resolution and image size (format) will not be as important as they would be if the unit were intended strictly for photography.

ACCESSORIES

Most manufacturers of starlight scopes also offer a variety of telephoto lenses for their product. Also, many models accept a C-mount or some similar adapter that allows one to utilize his present 35mm camera lens system with the scope in the same manner as he was able to do with the 16mm motion picture camera. The reader should refer to Chapter One, the section on "16 mm Motion Picture Cameras." Because in an urban area your camera-to-subject distance will generally be in the vicinity of 200 feet, a lens longer than 500mm will probably not be needed. If, however, it should ever become necessary to work at a distance of perhaps 300 to 1000 feet, a lens of 1000mm to 2000mm focal length will be essential to obtain an image size large enough to offer subject identification. A problem that could result if the amount of available light is very low is an image too dim to photograph effectively because long lenses are generally quite slow. A possible solution to this is to use an infrared radiation source with the scope, such as that discussed for infrared motion pictures. Such a radiation source will provide a very bright image even with a very slow lens. At the distance from which the radiation source would be used, there is very little chance the subject would see it and become curious about the dull red glow. One should remember that starlight scopes are very sensitive to the infrared region and will amplify that radiation just as they will the visible but weak radiation provided by a star. Also, when photographing with starlight scopes, fast black and white film must be used. Another point worth noting is that the image in the scope will affect the film differently than it will the light meter in the camera. If the manufacturer does not give specific instructions concerning this, it will be necessary to run exposure tests to determine at what ASA to set the camera's light meter to obtain correct exposures. This is nothing that will cause any difficulty. It is just something that one must be aware of and compensate for.

One innovation that some systems offer is a hinge back. The operator can mount both a biocular viewer and a camera attachment to the scope and use whichever he desires simply by swing-

Figure 45. Star-Tron model MK 202-A coupled with a biocular viewer (Photo courtesy of Smith & Wesson).

ing one to the side and the other into place. A setup such as this would make for easy viewing of the subject with both eyes by using the biocular viewer: When the action starts, one can quickly swing the viewer aside and the camera into place. For those who have sat and looked through a camera for eight or more hours, the advantage offered by this innovation should be obvious. It does not take long to develop fuzzy vision when looking through a camera for extended periods. Figure 45 shows a Star-Tron Model MK202-A® scope with a biocular viewer being used for general night observation.

Starlight scopes can be effectively used with 35mm single lens reflex cameras, motion picture cameras or closed circuit television cameras. Coupling starlight scopes to various cameras is easily done by utilizing a simple adapter which is mounted onto the scope in place of the eyepiece. The scope and adapter is then mounted to the camera body in much the same manner as a normal lens would be mounted.

This chapter was not by any means meant to be all inclusive but rather was intended to serve as a very general familiarization with what NVD's are and what can be done with them. While the many makes of starlight scopes are in general respects the same, they differ in detail and one should write to the various manufacturers or distributors for their promotional literature if he desires to make detailed comparisons.

CHAPTER NINE

HOW TO OPENLY PHOTOGRAPH A PERSON WITHOUT HIS KNOWLEDGE

Perhaps it will not happen often, but what does one do if he finds there is a need to photographically cover a person's activities but circumstances make it impossible to select a vantage point and work from a distance with telephoto lenses, and it becomes necessary to work within view of the subject? There are a number of simple techniques that make it possible to be quite close to a person, within his view, and to photograph him without his having any idea of the fact. The following techniques may serve to provide some ideas and while the reader may seldom, if ever, have occasion to use them, they may someday come in handy. He should consider also the interesting candid shots he might get at the next family reunion.

It is important that the camera be made ready for action so that, if and when the subject makes a sudden move, one does not lose the opportunity to record it because he was too busy focusing and determining the proper exposure. The camera should be all set so all you have to do is trip the shutter. To do this, one should cock the shutter, establish and set for proper exposure and check if from time to time; finally, one focuses hyperfocally if he does not know exactly where and how far away the subject will be when he suddenly has to make an exposure. All one does when focusing hyperfocally is note the f-setting of the lens and then study the depth of field scale on the lens barrel. From this it can be determined what the total range of acceptable focus will be and at what point perfect focus is. In other words, when the photographer focuses a lens on a subject, the point he focused on is in perfect focus and everything else from about half that distance to infinity will be in acceptable focus. This is, however, controlled to a great extent by the f-setting, assuming that the point of perfect focus is not just a few feet from the camera. For example, using a 50mm f1.4 lens on a Nikon camera (35mm) at f11, everything from about thirteen feet to infinity will be in acceptable focus. Whatever happens to be at about twenty-eight feet from the camera will be in perfect focus. With this kind of setting then, if the photographer saw some fast action take place and quickly released the shutter without focusing, he may not get a picture that is in perfect focus but he would, if the subject were not closer than the thirteen feet, get an acceptable photograph. If this technique were not used it is very likely the photographer would have no photograph at all. One should keep this in mind and practice it as it will make the photographer very quick on the draw, something that definitely draws a line between the pros and the amateurs.

Suppose that the subject and his family or friends should go to the fairgrounds, the zoo or the public picnic grounds and the photographer has seen fit to casually tag along with his *family.* In such settings he will not look conspicuous with his camera hanging around his neck. In some cases, perhaps he can keep his distance and photograph the subject with a 135mm or 200mm lens and the subject will not be aware of him. What does he do, however, if he suddenly finds that circumstances have put him at the picnic table next to his subject, perhaps not more than fifty feet away? The mild telephoto lens is now too long and a normal lens or perhaps even a mild wide-angle lens is necessary to get *good coverage of the whole set* because in most of the following techniques he will not use the viewfinder to frame the subject as that would serve to make the latter aware of the cameraman's interest in him. The photographer can try employing some of these techniques. If, for example, the subject is south of him, he can focus on something to the east or west that is about as far from him as the subject. After focusing and setting his exposure, he casually turns around 180° making it appear as though he was done with whatever he was photographing and wants to photograph something else. When he passes over the subject, he should trip the shutter and keep turning; the subject will never know.

Another technique is to have the exposure set and the focus set hyperfocally. One lets the camera hang in front of him with the strap around his neck. It is common that people with a camera in this position hold onto it and fondle it to some degree. Therefore, if one should be casually cocking the shutter and making exposures, someone fifty feet away would very likely have no idea of the fact. There is no need to look through the viewfinder if one has a normal lens on the camera as that offers plenty of coverage to allow for poor framing. If the subject is simply too close and the photographer is not sure of the framing, he should switch to a wide-angle lens. A variation of this is to have exposure and focus set as before, activate the self-timer and then set the camera down facing the subject while one lights a cigarette or whatever. No one would be suspicious of a camera just sitting alone unattended. The next technique is one that can effectively be applied to a Super 8 motion picture camera under conditions such as have been described. A 16mm camera would likely look out of place and arouse suspicion and curiosity in such a setting. If using a monopod, one should casually lean on it as it faces in the subject's direction. The photographer should appear to be interested in something else in a different direction, while watching the subject out of the corner of his eye. By using a cable release (it can also be done without one), he can be making a record of all the subject does without his being aware of it. This is nothing more than a variation of some of the above techniques applied to the 35mm camera. These ideas, with some imagination, should prove to be effective if one is ever in a like situation.

CHAPTER TEN

VANTAGE POINTS

THERE IS NOT MUCH that needs to be said concerning vantage points. The main thing is to secure some form of cover so as not to be *made* or observed by the subject, or to arouse sufficient curiosity or suspicion on the part of anyone else so that a report is made to the local law enforcement authorities. While they would do nothing more than possibly question the cameraman to establish what he was doing, that act would only serve to draw more attention to him.

Rooftops are generally good but care must be taken to avoid being silhouetted against the sky. Also one should avoid a rooftop directly across from the area to be observed. Another thing of importance, and this holds true for any vantage point, is that a discreet means of coming and going and shift changes must be established. If circumstances permit, a hotel room, apartment, or a business establishment can be secured, and photographing done from that point. If it is summer, the window can often be opened and photographs taken through the opening. If, however, it is winter, make every effort to photograph through not more than one pane of glass and preferably a thick pane as opposed to the thin panes commonly used in private homes. The problem encountered in photographing through glass is that it refracts the light and an unsharp and distorted image can result. The thick glass used on store fronts, and so forth, tends to refract less than the thin panes used in homes as they tend to be more uniform in thickness. It is also extremely important that the glass be very clean, since dirty windows will cause the resulting photographs to be unsharp. If the room is darkened, one will encounter fewer problems with reflections. Finally, if the glass one is photographing through is causing an unsharp image, the problem will become worse as you go to lenses with longer focal lengths. If the windows have curtains, shades, and so forth, one should try to set up so he can observe and photograph from the top of the curtain rather than through the curtain by parting it a few inches. In some instances a person will be able to work from a darkened room with the curtains partially or fully open. This is possible especially if working from a second- or third-story room. Trucks are very effective as they can be used in more situations than not. A panel truck or pickup truck with a shell can easily be equipped with one-way glass windows or soap can be smeared on the inside of the window with a peephole left. Also worth consideration is a fictitious business name painted on the side or rear windows of such a vehicle with the inside of a letter or letters left unpainted through which observation and/or photographing can be done. If forced to work from a passenger car temporarily, one should sit in the back seat. One should also consider raising the head rests on the front seats and lowering the visors. When working from a passenger car, if cost and circumstances permit it, the photographer ought to work with a female investigator as a couple sitting in a car will generally arouse little suspicion. Also every effort should be made to avoid parking on the same block as the subject, since the chances of being *made* will be greatly reduced if one can work from the next block.

While this discussion of vantage points is not and cannot be all inclusive, it will be found that most vantage points are nothing more than variations of what has been discussed. The reader is referred to the chapter dealing with telephoto photography as it discusses a number of things that should be taken into consideration from a photographic standpoint when selecting a vantage point.

CHAPTER ELEVEN

SURVEILLANCE PHOTOGRAPHS AND THE LAW

General Considerations

G ENERALLY THE RULES of evidence as they apply to the collection, handling and introduction of any kind of physical evidence, as well as crime scene photographs, will apply to surveillance photographs. The most important thing is a good solid chain of custody showing that the history of the evidence can be accurately accounted for. It is also essential that the photographs be relevant to the case and in no way misleading. Photographs to be used as evidence must be a true representation of what was photographed and must be free from any distortions. The photographer must be in a position to verify that the films or prints are a true resemblance of what he observed and photographed.

CHAIN OF EVIDENCE

In situations where the services of a commercial laboratory are utilized for the processing of color and/or motion picture films, and so forth, it is essential to have an affidavit showing who received the films, who supervised the processing, and in general, who can account for the materials from the time they were given to the laboratory until the cameraman signed for them and took them with him. If the laboratory service he wishes to do business with does not have an affidavit for this purpose, he can make some up himself. The affidavit shown here is the form used by Sly-Fox Films, Inc., located in Minneapolis, Minnesota to maintain a chain of custody. Also, one must be sure to instruct the laboratory personnel to keep the film intact. No part of the film may be cut off no matter how unimportant it may appear to be. This holds true even if a portion of the film is blank. Similarly, nothing may be added to the film. After signing for and receiving the processed materials from the laboratory, if he intends to pass them on to someone else, the photographer should be sure a receipt is made out and signed to show when and to whom the evidence was given. These are basic rules that apply to evidence in general. They also apply to photographic evidence.

AFFIDAVIT*

STATE OF MINNESOTA ⎫
COUNTY OF HENNEPIN ⎰ ss.

I, ..., having been duly sworn, on oath, depose and say that I am the .... of Sly-Fox Films, Inc., 1025 Currie Avenue, Minneapolis, Minnesota 55403.

That on the .... day of ...., 19...., at ...... o'clock ....M. I received from .... at the office of Sly-Fox Films, Inc., as shown above,

of film for development or printing.

That the films in the said containers were processed under my supervision and control and they were processed in the normal and customary manner for development of film of the particular type.

That these films were not cut, edited, changed, negatives were not reversed, superimposed by other film, retouched, over or under developed or was any other thing done which could alter or change the film in any manner or what the film attempts to portray.

That after processing the films and/or prints were returned to .... at .... o'clock ....M. on the ...... day of ...., 19.....

....
Subscribed and sworn to before me this ...... day of ...., 19.....
....

VALIDITY OF EVIDENCE

Before attempting to introduce your films or photographs as evidence in a court of law, be sure the films or photographs are a true and accurate representation of the scene or whatever it is they are intended to establish. Also, anything one can obtain to support photographic evidence will be helpful. One should consider making a sketch showing where the subject was at the time he was photographed, where the camera and photographer were, possibly the position of the sun as well as anything else that may be pertinent under the circumstances. Some surveillance photographers when shooting motion pictures like to expose a few feet

---

* Courtesy of Sly-Fox Films, Inc.

at the beginning of the film showing the front page of that day's newspaper with the date, and so forth. While this cannot prove that the filming was not made at some later date, it will serve to establish that it was not made prior to that date. Also, if there is anyone besides the photographer who can testify to such things as when, where and how your photographic evidence was obtained, that is also to your advantage.

TRESPASSING

When engaged in photographic surveillance, one will do well to avoid trespassing on a subject's property in an effort to obtain photographs. To trespass is to open one's self and one's case up to attack once in the courtroom, and possibly before. This is more important in civil cases than in criminal. If at night a subject is in his home with the drapes open and lights on, thus making it possible to photograph him from the street, the cameraman is not trespassing nor is he violating his right to privacy by photographing the subject from the street as he has no reasonable expectation of privacy in that situation. Should the drapes be closed, however, and should one discover upon walking into his yard and up to the window that the drapes have a gap of an inch or so and should one take photographs through the gap, he has indeed trespassed by walking onto the property and the court will in all likelihood consider that he has invaded the subject's right to privacy stating that he (the subject) had a reasonable expectation of privacy with his drapes closed. Some states will allow one to go onto the subject's property but where this can be avoided it is desirable to do so.

RECORDS AND FILING

After securing the desired photographic evidence, it is important that films, negatives, prints, and so on, be properly filed. Although it is not usually necessary to include all the technical information such as camera settings, and other details on the back of each individual print, the file or case number on the back of the prints should direct one to the appropriate record in the file bearing all pertinent information. The method of filing will vary with various organizations and whatever method you choose is not important so long as it is accurate. It is absolutely essential to have files under lock and key to avoid having to admit in court that it would not be difficult for someone to get in and tamper with the contents. With photographic evidence, just as with any kind of evidence, if it can be discredited or if enough doubt can be cast upon it, the photographs will not be allowed in as evidence. Accuracy and security of the files is important!

The following is a basic summary of what has been discussed:

1. Maintain a solid chain of evidence.
2. Obtain as much evidence as possible to support photographic evidence.
3. Obtain the evidence in a manner that does not leave one open to reproach.
4. Be sure the evidence is pertinent to the case and not just meant to be sensational.
5. Maintain accurate and secure records and files.
6. Do not cut or splice (alter) motion picture films in any manner.

INDEX

A

Aberrations, spherical, 30
Accessories, 23, 86-88
Adapters
  c-mount, 7, 86
  hinge back, 87
  binocular, 16
Affidavit, 94-95
Agitation, film development, 62
Air turbulence, 34-36
Aircraft landing light for infrared radia-
  tion, 78
Altering motion picture film, 95, 97
Atmosphere, 34-36
Automatic lenses, 22-23, 28
Auxiliary focusing mark, infrared, 65-
  66

B

Back lighting, 35
Baffles, light, 27
BCPS (Beam Candle Power Seconds),
  64, 78
Beam splitter prisms, 6
Beanbag, 42-44
Beaulieu, motion picture camera, 7
Bell & Howell, motion picture camera,
  7-8, 11-12
Belt-Pod, 39-40
Binoculars, 46
  adapter for cameras, 16
Biocular viewer for NVD's, 87-88
Bolex, motion picture camera, 5-6

C

Cable release, 43
Camera types for surveillance photog-
  raphy, 3-16
Camron lenses, 23
Candlepower (BCPS), 64, 78

Cassette for film, 6
Chain of evidence, 94-95, 97
C-Mount adapter, 7, 86
Compound refractive lens, 23-27
Cutting motion picture film, 95, 97

D

Density of film, 60
Depth of field scale, 89-90
Developing formulas for push-process-
  ing films, 62
Developing kit, E-4, 61
Diffusion
  glass, 68
  image, 36
Direct objective lens, 17-18, 23-27

E

E-4 developing kit, 61
Ektachrome EF, Kodak, 10, 48, 54, 61
Ektachrome-X, Kodak 61
Electric sequence unit, 5
Electromagnetic radiation, 64
Electron tubes, 80, 82, 84
Evidence
  chain of, 94-95
  receipt of, 94-95
  rules of, 94-95
  validity of, 94-95
Eposure determination for infrared
  photography, 66-68, 78
Eyepiece magnifier, 47, 57

F

Files, security of, 97
Filing, records, 96
Films for night photography, 9-10, 62
  (see also specific type of film, development of) (see Force-Processing, 51, 58-63, 68)

Filters
  Kodak Wratten (infrared), 64-65, 68, 72, 77
  neutral density, 21-22
First generation starlight scopes, 84-86
Flare, lens, 36
Flash bulbs, infrared, 66, 75
Flock paper, 27
Focusing adjustment, infrared, 65-67
Focusing hyperfocally, 89-91
Force-processing films, 51, 58-63
  infrared film, 68
Formulas for push-processing films, 62
Fresnel lens, 67-68
Front lighting, 35

G

Gadget bag for tripod weight, 45
Gafpan high-speed reversal film, 10
Gelatin filters, 64-65
Glass, one-way, 54, 93
Guide numbers, 66-68, 78
Gun stock mount, 41-42

H

High speed Ektachrome, Kodak, 48, 54, 61-62
Honeywell Penta
  infrared camera, Nocta, 73-75
  1°21° light (spot) meter, 49-51
Hood, lens, 36
Hot spots, infrared photography, 68
Hyperfocal, focus, 89-91

I

Image diffusion, 36
  tube, 80, 84-86
Infrared
  auxiliary focusing mark, 65-67
  camera, 73-75
  film, 64-65
  filters, 64-65, 68, 72, 77
  flash bulbs, 66, 75
  guide numbers, 66-68, 78
  photography, 64-79
  push-processing infrared film, 68
Intensification, light, 81-82
Independent lens manufacturers, 22-23

J

Javelin, starlight scope, 84-85

K

Keystone motion picture camera, 10
Keystone Cop Effect, 8
Kodak 2475 high-speed recording film, 51, 55, 62

L

Landing light, aircraft for infrared radiation, 78
Lens and camera shake, 36-47
Lens manufacturers, major and independent, 22-23
Light
  baffles, 27
  intensification, 81-82
  meters
    Honeywell Pentax 1°21°, 49-51
    Minolta, 49
    Soligor, 49
  scatter, 34-36
  spectrum, 64
Lighting
  back, 35, 57
  front, 35, 57
  scattering, 35, 57
  side, 35, 57
Long focus lenses, 17-18, 23-27

M

Magnifier, eyepiece, 47, 57
Major lens manufacturers, 22-23
Minolta
  subminiature camera, 13-16
  light meter, 1°21°, 49
Minox subminiature camera, 13
Mirror lock up on 35mm single lens reflex cameras, 46
Mirrors, lenses, 17, 19-27, 37, 67
Mobile unit, surveillance truck, 54
Monopod, 40, 91
Motion picture cameras, 16mm vs. super 8, 10-11
Motor drive units, 4, 13
Movement, lens and camera, 36-47

N

Neutral density filters, 21-22
Night vision devices (NVD's), 72, 80-88
Nikon, 4, 20-22
Nocta, Honeywell Pentax infrared camera, 73-75
Normal lens, usefulness of, 54

O

One-way glass, 54, 93
Openly photographing people without their knowledge, 89-91

P

Paper, flock, 27
Parabolic reflector (Infrared Flash Unit), 75
Parallax, 12, 73
Perspective, telephoto lenses, 17
Photons, 81
Pocket cameras, Kodak, 13-14
Preset lenses, 22, 28
Privacy, reasonable expectation of, 96
Push-processing film, 51, 57-63
  infrared film, 68

Q

Questar, 34

R

Radiation, electromagnetic, 64
  sources for instantaneous and continuous infrared exposures, 64-73, 86
Radio control unit, 4
Range finder cameras, 3, 8, 11
Receipt of evidence, 95
Reciprocity effect, 63
Recording film, 9, 51, 55, 62
Records
  filing, 96-97
  security, 97
Reflector, parabolic for infrared flash, 75
Reflex cameras, 3-4, 11-12

Reflex-mirror lens, 19-22, 37, 67
Refractive lenses, 17-27, 37, 67
Reversal film
  4-X, 9
  Gafpan, 10
Right to privacy, reasonable expectation of, 96
Roll film, 6
Rules of evidence, 94

S

Sandbag for camera support, 42-44
Safety margin, film speed, 59
Second generation starlight scopes, 84-86
Security of files and records, 97
Self-Timer, 46, 91
Sequence exposures, 3-5
Shake of lens and camera, 36-47
Shutters, motion picture cameras, 7-8
Side lighting, telephotography, 35
Silhouette from back lighting, 57
Single-frame release, 4
Single lens reflex cameras, 4, 11-12, 58
Sketch of camera and subject location, 95
Slow motion, motion pictures, 5
Sly-Fox Films, Inc., 94-95
Sniperscopes, 72-75, 80-84
Snooperscopes, 72-75, 80-84
Soligor
  lenses, 23
  light meter, 49
Spectrum, light, 64
Spherical aberrations, 30
Splicing motion picture film, 95, 97
Starlight scopes, 72, 80-88
Star-Tron, 83, 87-88
Subminiature cameras, 3, 13-16
Super 8 motion picture cameras, 6, 10, 17, 91
Supports, camera, 36-45
Surveillance truck with one-way glass, 54

T

Tamron, 23
Tele-Extenders, 11, 27-34

Telephoto flash, 67
Telephoto lenses, true, 18-22
Telephotography, 17-47
Telescope building, 27
Tessina, subminiature camera, 13
Timer, Self, 46, 91
Trespassing, 96
Tripods, 38, 44-45
Tri-X film, 51, 62
   photos taken with, 25-26, 52-53, 56,
     85
Trucks, surveillance, 54, 93
Turbulence, air, 34-36

Turret for filters on mirror lenses, 21-
   22

V

Validity of evidence, 95-96
Vantage points, 35, 92-93
Vibrations of lens and camera, 36-47
Vignetting of image with mirror lenses,
   22
Vivitar, 23, 29, 34, 40

W

Wide-Angle lens, 91
Window mount, 40-41